CAMBRIDGE MONOGRAPHS IN
EXPERIMENTAL BIOLOGY
No. 15

EDITORS:

P. W. BRIAN, G. M. HUGHES
GEORGE SALT (*General Editor*)
E. N. WILLMER

PROTONS, ELECTRONS, PHOSPHORYLATION AND ACTIVE TRANSPORT

THE SERIES

PROTONS, ELECTRONS, PHOSPHORYLATION AND ACTIVE TRANSPORT

BY

R.N.ROBERTSON, F.R.S.

Professor of Botany in the University of Adelaide

CAMBRIDGE

AT THE UNIVERSITY PRESS

1970

Published by the Syndics of the Cambridge University Press
Bentley House, 200 Euston Road, London, N.W.1
American Branch: 32 East 57th Street, New York, N.Y.10022

Library of Congress Catalogue Card Number: 68–26988

ISBN 0 521 07324 3

First published 1968
Reprinted 1970

First printed in Great Britain at the University Printing House, Cambridge
Reprinted in Great Britain by Lewis Reprints Limited, Port Talbot, Glamorgan

CONTENTS

v

PREFACE

THIS book puts a personal point of view—that the separation of protons and electrons from hydrogen atoms of water or organic molecules is a very important fundamental biological process. The book is not a comprehensive treatise and, far from considering all the relevant literature, will draw on a limited number of experiments and observations which illustrate the principles; it is intended for physiologists and biochemists and for biologists with some general knowledge of biochemistry. Most but not all the examples are from work on plants but the principles, I believe, apply to a wide range of organisms. For this reason I hope that zoologists and microbiologists may find the book useful.

This work is based on a series of five lectures given while I was a visitor at the Botany Department, Imperial College, London. I wish to thank that Department and particularly Professor W. O. James, F.R.S., and Professor C. P. Whittingham for their hospitality. I benefited greatly from discussions with staff and research workers and also with those from other London departments, from the University of Cambridge, and from the University of East Anglia, who attended the lectures. I am particularly grateful to Dr E. A. C. MacRobbie (Cambridge) and to Dr J. T. Wiskich (Adelaide), who read much of the manuscript and assisted with critical comments. Not surprisingly, this book is rather different from the lectures.

I also wish to thank my secretary, Miss Carolyn Wilkins, for her continued help and for typing the manuscript.

R. N. ROBERTSON

November, 1967

vii

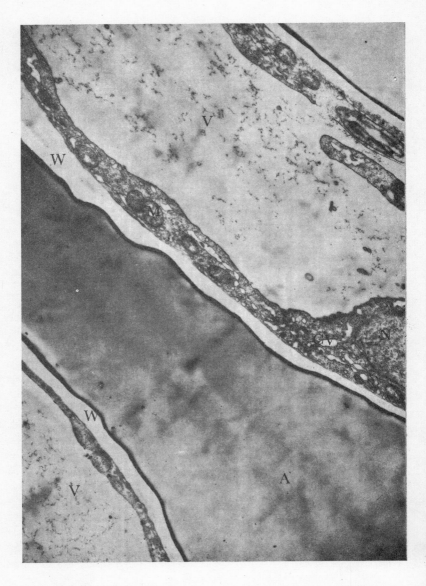

Epidermal cell and bladder cell of *Atriplex spongiosa*. The bladder cell (at the top) accumulates sodium chloride in the vacuole but the epidermal cell (bottom left) does not. Notice the many organelles and vesicles in the cytoplasm of the bladder cell, in contrast to the cytoplasm of the epidermal cell. (C. B. Osmond and K. R. West, personal communication.) V, vacuole; N, nucleus; W, wall; M, mitochondrion; CV, cytoplasmic vesicles; A, air space. × 10000.

CHAPTER I

A Hypothesis was Developed...

'...the vision of those workers in ion transport whose speculations first led in this direction has been experimentally
justified'. (JAGENDORF AND URIBE)

LIVING matter uses energy to make the normal physical
processes of non-living matter appear to run backwards. If
we use the analogy of water running downhill turning a wheel
from which energy is obtained, living matter appears to make
the wheel run backwards and move water uphill so its potential
energy is increased. Like all analogies this is imperfect, and
others have described the phenomenon in different ways.
Schrödinger (1948) in *What is Life?* says that 'living matter
feeds on negative entropy, i.e. unlike non-living matter,
extracts order from its environment'. The simplest way of
describing what happens is to say that living matter uses energy
arriving at the earth's surface to split water into oxygen, which
is liberated, and protons and electrons and, in the multi-
reaction process of photosynthesis, to transfer those electrons
and protons to carbon dioxide, which becomes carbohydrate.
As long as they are associated with carbon, these hydrogens
are a potential source of energy and on them all living processes
depend. The electrons and protons are shuffled round in
various ways, sometimes together, sometimes apart, but always
capable of bringing about a reaction or an energy-requiring
physical process until the final point where they are passed
back to an oxygen atom to form water again, after the carbon
is eliminated as carbon dioxide. At that point no more energy
can be derived from that water or carbon dioxide—and light
energy must be brought in again.

How are the reactions, starting with those of photosynthesis
and ending with those of respiration, linked to such diverse
living processes as synthesis of complex molecules, active
transport of ions and molecules, maintenance of bioelectric and

I

bioluminescent processes? In this book, the idea that the separation of a positive and a negative charge in both respiration and photosynthesis is fundamental, is developed in the light of recent work. We have become increasingly aware that the following characteristics are associated with biological membranes which are usually no more than 50–70 Å in thickness: (1) very low permeability to ions, (2) built-in enzyme systems which conduct protons to one side of the membrane and electrons to the other, (3) a source of hydrogen atoms from which protons and electrons are derived, (4) specific carriers or enzymes which react differently according to conditions on the two sides of the membranes.

Despite the very long period in which bioelectric phenomena have been known, the mechanism which separates a positive and a negative charge in both respiration and photosynthesis had not been studied until recently. About 40 years ago, Lund (1928) suggested that the sources of bioelectric phenomena might be related to the oxidation–reduction reactions characteristic of respiration. The idea attracted little attention for the only material across which separation of positive and negative charges might be maintained would be the predominantly lipid membranes, which do not seem very likely to bring about the separation in the first place. Most living systems appear to be predominantly aqueous media in which charges are carried by ions. Ten years later, Friedenwald and Stiehler (1938) and Stiehler and Flexner (1938) used the idea that movement of ions and secretory processes might depend on oxidation–reduction, but the important development was due to Lundegårdh (1939) with his publication of an electrochemical theory of salt absorption and respiration. Lundegårdh and Burström (1933, 1935) had shown that both salt uptake and oxygen uptake by roots were inhibited by cyanide. Lundegårdh realized that when the cytochrome system, of which the terminal member is inhibited by cyanide, was reduced by accepting an electron into the iron atom of the porphyrin, a proton was probably released simultaneously into the medium. He suggested that this separation of charge might result in hydrogen ions leaving the cell by one pathway and hydroxyl ions leaving the cell by another. He suggested that a cation could enter by exchange for a hydrogen ion and that an anion could enter by exchange for a hydroxyl ion; salts

2

would accumulate inside the cell and water would be formed outside. In his expansion of the hypothesis, Lundegårdh (1940) included the idea that electrons might move in an appropriate system (such as an electron-conducting system) without actual movement of the electron carrier. Lundegårdh's hypothesis, modified from time to time, is probably best stated in its early form in the paper of 1945 on absorption, transport and exudation of inorganic ions by roots, but the history of his hypothesis is outlined in his book on plant physiology (Lundegårdh, 1966).

Though Lundegårdh had been concerned with quantitative experiments, he had not attempted a stoichiometric analysis. In 1945, I began experiments to investigate the stoichiometry of salt respiration and salt accumulation. In speculating on why salt respiration, which is the stimulated respiration found in Lundegårdh's roots and in my experimental material (carrot disks), seemed to bring about salt accumulation, and why both were stopped by cyanide, I thought that the cytochrome might not only be the site of separation of protons and electrons but also be an anion carrier. If cytochrome could take an electron (unit negative charge) in one direction, it might take an anion (unit negative charge) in the other. The same idea had previously occurred to Lundegårdh (1945), as I learned when his paper became available. The fate of this hypothesis will become clear later in the book.

The quantitative idea was important. If the hypothesis were correct, the maximum rate at which such a system could operate would be when one electron carried out was balanced by one monovalent anion carried in. The maximum rate would be predictable from the amount of oxygen taken up in respiration.

Since $C_6H_{12}O_6 + 6H_2O + 6O_2 \rightarrow 6CO_2 + 12H_2O,$

it follows that separation of protons and electrons for each $6O_2$ would give $24H^+$ and $24e^-$ so that if an anion entered when an electron passed out through the carrier and a cation exchanged for a proton, 24 monovalent anions and 24 monovalent cations would be accumulated; that is, at the maximum rate,

$$\frac{\text{g equiv. salt absorbed}}{\text{g mol. oxygen absorbed}} = 4.$$

The data of my previous work were not adequate to establish the stoichiometry since they contained many experiments in which the rate of salt uptake was limited by salt concentration, but they showed that the amounts were of the right order of magnitude. During the next three years, Miss Wilkins and I carried out a series of experiments on a variety of salts, measuring simultaneously the salt absorption and the salt-stimulated respiration in carrot tissue, leading to our publications in 1948 (*a* and *b*). This experimental material showed rates of salt accumulation which were consistent with one anion entering when one electron passed out to the oxygen through the carriers and one cation entering in exchange for the hydrogen ion simultaneously passing by a different path to the oxygen.

While we were investigating the quantitative relations between salt accumulation and salt respiration in carrot tissue, various people were looking at the quantitative relations between the stimulated oxygen uptake of the gastric mucosa and the amount of hydrochloric acid secreted simultaneously (Conway and Brady, 1948; Crane and Davies, 1948*a*, *b*; Patterson and Stetten, 1949). Though Conway and Brady published the first suggestion that protons released at the cytochrome system might be the source of the hydrogen ions of the hydrochloric acid, the first satisfactory measurements were those of Crane and Davies. An equivalent amount of alkali is known to pass back to the blood as bicarbonate ions resulting from the action of the enzyme carbonic anhydrase on hydroxyl ions and carbon dioxide. The history of these investigations was discussed by Davies (1951).

By the end of 1948 a hypothesis connecting both hydrogen ion production in the gastric mucosa and accumulation of ions by plant cells with the proton and electron separation was supported by the quantitative evidence. The principles are illustrated in fig. 1. No attempt is made in this diagram to identify the membranes concerned—these will be discussed later in the book. The quantitative relation of the oxygen uptake in respiration to the amounts of acid secreted (gastric mucosa) or of salt accumulated (plant cells) would be as follows:

Stage I. Respiration $C_6H_{12}O_6 + 6H_2O \rightarrow 6CO_2 + 24H^+ + 24e^-$
$$\begin{vmatrix} +6O_2 \\ \downarrow +12H_2O \\ 24OH^- \end{vmatrix}$$

Stage II. HCl secretion $24H^+$, $24Cl^-$ (left)/$24HCO_3^-$, $24Na^+$ (right).
Stage III. NaCl accumulation $24Na^+$, $24Cl^-$ (left)/$12H_2O$ (right).
Ratio: g mol. secreted or accumulated to g mol. O_2 absorbed $= 4$.

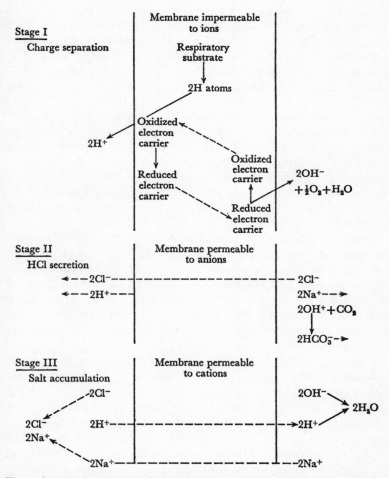

Fig. 1. Stage I: the principle of separation of protons and electrons in a membrane due to the orientation of the electron carrier. Stage II: chloride movement from right to left (e.g. blood side to secretory side in gastric mucosa) with an anion permeable stage in the membrane. Stage III: sodium chloride accumulation on the left (e.g. in a plant cell) if the membrane were permeable to hydrogen ions and cations after chloride had moved; water formed on the right. Solid arrows, chemical reactions; broken arrows, diffusion pathways.

5

In the years that followed, various experiments suggested that the hypothesis was not as simple as that outlined in fig. 1 and was not dependent on the cytochrome separation of protons and electrons alone. Crane and Davies (1951) found evidence that the ratio of HCl secreted may exceed 4 and be as high as 12. This high ratio has been disputed by Davenport and his collaborators and was still a matter of controversy in 1957 when both Davies (1957) and Davenport (1957) outlined conflicting evidence which led Davies to conclude that the ratio was 12 and Davenport to conclude that it did not exceed 4. Conway (1959) and Robertson (1960) concluded that there was no good evidence that the ratio exceeded 4.

The belief that the ratio might be 12, right or wrong, has important historical interest for it led Davies and Ogston (1950) and Davies and Krebs (1952) to postulate that some additional hydrogen atoms, and hence ions and electrons, might also be derived from the phosphate carriers. This was the first suggestion, further developed by Davies (1961), of an interdependent relationship between the separation of charge at the electron carrier and the phosphorylation.

Another apparent complication in the hypothesis came from the work on 2,4-dinitrophenol (2,4-DNP) which, as Loomis and Lipman (1948) had shown, uncouples phosphorylation from oxidative respiration, i.e. prevents the formation of adenosine triphosphate (ATP) from adenosine diphosphate (ADP) and inorganic phosphate (P_i), without inhibiting the simultaneous oxygen uptake. The same substance inhibited salt accumulation in carrot cells (Robertson, Wilkins and Weeks, 1951) and hydrochloric acid secretion by the gastric mucosa (Davies, 1951). Our work on carrot tissue showed that the oxygen uptake in the presence of 2,4-DNP was increased and proceeded via the cytochrome system but did not support ion accumulation. Did this mean that the processes of salt accumulation and hydrochloric acid secretion were dependent on ATP formation inhibited by 2,4-DNP—or that both the active transport processes and the ATP formation depended on a common cause which was inhibited by 2,4-DNP?

The other important development was our increasing awareness of the role of mitochondria. When our hypothesis of the role of cytochrome was developed, we knew little about the occurrence of cytochromes in the cell except that they were associated with particulate material. From 1948 onwards, thanks to the work of Lehninger, Green, Hogeboom and others, on isolated mitochondria (see Lehninger, 1964), we realized that the cytochrome system was firmly attached to the mitochondria. The first work on plant mitochondria isolated for this type of experiment was published by Du Buy, Woods and Lackey in 1950. By 1960 many of the properties of mitochondria, both plant and animal, were well known and may be summarized:

(1) They are recognizable structures or organelles within cells with a smooth outer membrane and a much convoluted inner membrane.

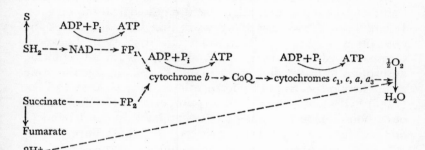

Fig. 2. The electron transport chain; protons and electrons move independently. Solid arrows, chemical reactions; broken arrows, pathways. FP, Flavoprotein; NAD, nicotinamide adenine dinucleotide; CoQ, ubiquinone.

(2) They contain the dehydrogenases which catalyse the loss of hydrogens from the substrate molecules of the Krebs acid cycle.

(3) Their membranes contain the electron transport chain which results in the passage of electrons from the substrate hydrogens to oxygen, where, with the acquisition of protons, water is formed (fig. 2). Electrons and protons result from the dehydrogenation of substrates and the electrons are passed from carrier to carrier, the protons often being dissociated. There is still some uncertainty about the details of the chain. For example, the exact place and the role of ubiquinone is not certain and it is not clearly established what parts are played by the non-haem iron, the sulphydryl groups or the copper which is associated with the cytochrome oxidase.

B

(4) At certain places in the chain the formation of adenosine triphosphate (ATP) from adenosine diphosphate (ADP) and inorganic phosphate (P_i) must accompany the oxidation–reduction. If phosphorylation does not occur, electrons cannot be transferred and oxygen uptake will not occur either. There is compulsory coupling between electron transport and oxidative phosphorylation.

(5) The uncouplers such as 2,4-DNP will allow electron transport, oxygen uptake and water formation without concurrent phosphorylation. As we have seen, uncouplers inhibit the secretion of hydrochloric acid and the accumulation of ions in plant cells.

By 1960 there was evidence from an entirely different source that a charge separation might be of physiological importance. It had been known for many years that light stimulates the absorption of salts by green cells. In 1957, Van Lookeren Campagne showed that the action spectrum of chloride absorption of leaves of a water plant, *Vallisneria*, was the same as the action spectrum of photosynthesis, but, as Arisz and Sol (1956) had shown, chloride absorption occurred in the absence of carbon dioxide and was therefore apparently not dependent on the full photosynthetic process. In 1963 Bové, Bové, Whatley and Arnon showed that chloride was essential for certain steps in photosynthesis at which a transfer of electrons, with probable separation of electrons and protons, was postulated. This raised the possibility that charge separation occurring in chloroplasts in the light might be basically responsible for the consequent ion movements.

In 1960 I reviewed the situation and attempted to reconcile our knowledge of respiration, oxidative phosphorylation and active transport of ions, and some of the problems reviewed at that time will be useful in introducing the important chemiosmotic hypothesis of Mitchell.

CHAPTER 2

And Extended...the Mitchell Hypothesis

'Research is to see what everybody has seen and think what
nobody has thought'.   (SZENT-GYÖRGI)

THE processes which result in secretion of ions as hydrochloric
acid in the stomach or accumulation of ions in the vacuoles of
plant cells are termed 'active transport' mechanisms. 'Active
transport' was a term used by Ussing in 1949 and clearly
defined by him (1949, 1957) as a process whereby an ion is
moved against its electrochemical potential gradient; active
transport is therefore dependent upon energy produced by the
cell. If one ion, either cation or anion, is moved against the
electrochemical potential gradient in secretion or accumulation, that ion has been actively transported. The mobile ion
that will then move with it may be moving against its concentration gradient but with its electrochemical potential
gradient to maintain electrical neutrality, so both ions of a salt
will be secreted or accumulated. Only one ion need be moved
in the first place with expenditure of energy on the part of the
cell. The definition which has been accepted generally by
physiologists and biochemists makes no assumptions about how
the energy is provided to move an ion in the first place, and
thereby differs from the more rigorous (and more restrictive)
use of the term 'active transport' by Spanner (1964).

Active transport systems are of different kinds and the danger
of generalizing too widely must be avoided. The early development of the hypothesis arose, as we have seen, from hydrochloric acid secretion and salt accumulation, both aerobic
processes. Other kinds of active transport (e.g. in mammalian
erythrocytes) are anaerobic, and some cells (e.g. frog skin)
have both aerobic and anaerobic mechanisms. Many cells
have a sodium–potassium pump which extrudes sodium and
takes up potassium by an active process now known to depend
on the breakdown of ATP by an ATP-ase. Early speculations

9

on active transport were often based on the assumption that mechanisms must all depend on the provision of energy from ATP, that ubiquitous energy carrier. One purpose of my review in 1960 was to point out that some active transports,

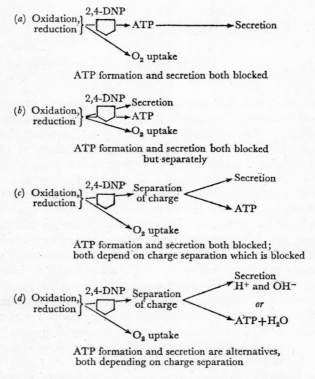

(a) Oxidation, reduction → 2,4-DNP → ATP ————→ Secretion

O₂ uptake

ATP formation and secretion both blocked

(b) Oxidation, reduction → 2,4-DNP → Secretion / ATP / O₂ uptake

ATP formation and secretion both blocked but separately

(c) Oxidation, reduction → 2,4-DNP → Separation of charge → Secretion / ATP

O₂ uptake

ATP formation and secretion both blocked; both depend on charge separation which is blocked

(d) Oxidation, reduction → 2,4-DNP → Separation of charge → Secretion H⁺ and OH⁻ or ATP+H₂O

O₂ uptake

ATP formation and secretion are alternatives, both depending on charge separation

Fig. 3. Possible interpretations of the observation that uncouplers like 2,4-dinitrophenol (2,4-DNP) allow oxygen uptake but block both phosphorylation and active transport or secretion (Robertson, 1960).

oxidative phosphorylations, charge separations, and oxygen uptakes might all be closely involved with each other and that the initial causative relation required further investigation. The evidence was that uncouplers, which prevent oxidative phosphorylation and ATP formation without stopping oxygen uptake, also prevent active transport in HCl secretion and salt accumulation. The possible interpretations given then are set out in fig. 3. If all secretion mechanisms were similar to that of the sodium–potassium pump, already investigated by Caldwell (1956) and Caldwell and Keynes (1957), the first interpretation would apply. In the absence of more definitive evidence than existed then, any of the other interpretations would have been possible for other forms of active transport. The principal contribution of my review in 1960 was probably to emphasize that, as far as we knew, the *separation of positive and negative charge might be a fundamental process in electron transport (whether in mitochondrial respiration or chloroplast photosynthesis) which preceded the movement of ions and the phosphorylation of ADP to give ATP.* Further, it was suggested that the two consequences of charge separation—that is, secretion of ions or phosphorylation—might be alternatives. As we shall see, this idea has been justified by subsequent hypothesis and experiment. In retrospect, it is interesting to note that I did not have enough insight to suggest that the breakdown of ATP might bring about a charge separation and hence an ion movement —a possibility which Mitchell must have seen clearly at about this time.

Before going on to consider Mitchell's hypothesis, which made the really important advances at this time, one other topic of my 1960 review should be discussed here because it has important bearing on the development of current theory. The hypothesis was that the hydrogen ions of hydrochloric acid were really the protons liberated from the respiratory process when the electrons were separated from them in the electron transport chain. At first sight, and indeed in earlier hypotheses, the likely explanation would be that the hydrogen ions are left on one side of a membrane when the electrons move back from them, either through an electron-conducting system of conjugated double bonds or by a mobile electron carrier. But there are two difficulties: first, the concentration of chloride (166 m-equiv./l.) secreted exceeds that of the hydrogen ion (159 m-equiv./l.), the difference being made up by potassium (7 m-equiv./l.) (Gray, 1943), which would be impossible if the chloride were following the potential set up by secreted hydrogen ions; secondly, the potential difference between the secretory side and the blood side would be the wrong way round if the chloride were following the hydrogen. Conclusive experiments by Hogben (1951, 1955) with tracer fluxes on short-circuited mucosae showed that the chloride was being actively transported. Thus it becomes necessary to postulate that the chloride ion is actually removed from the solution on one side and transported to the other side of the membrane. On the hypothesis originally stated by Robertson and Wilkins (1948a, b), this would imply combination with an anion carrier which might be the oxidized electron carrier with net positive charge (fig. 4). The transported anion would be liberated

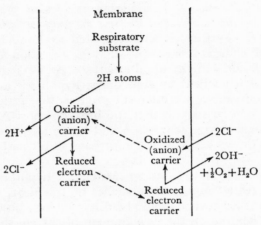

Fig. 4. Principle of an electron carrier in one direction acting as an anion carrier in the opposite direction; active transport of an anion or an anion pump.

from the carrier when it became reduced by the next electron. A mechanism of this kind would be more consistent with anion transport as the primary act of secretion than the alternative possibility that electrons, without simultaneous anion transport, moved from the protons through an electron-conducting system. Lundegårdh too has always believed in the necessity for anion transport to accompany electron transport (see Lundegårdh, 1966). We shall return to the possible importance of this concept in discussing the charge separation in organelles (chapter 4).

Mitchell's chemi-osmotic hypothesis

Great credit must be given to Mitchell who, seeing what other people see, thought 'what nobody had thought'. He had long been concerned with the role of membranes which not only separate substances but also contain reactive molecules such as enzymes. Since the orientation of an enzyme in a membrane can confer directional properties on the reactions involved, Mitchell had postulated that the results of a reaction may be different on the inside from those on the outside of a membrane. The original hypothesis was outlined in 1961 and, with a number of modifications and additions, was stated in comprehensive form in 1966 (Mitchell, 1961, 1962, 1966). He accepted the idea that separation of charge occurs in normal mitochondria and chloroplasts and is responsible for the formation of ATP from ADP and P_i. His first important new suggestion was that this synthesis might be due to the reversal of an ATP-ase reaction. An ATP-ase which catalyses the hydrolysis of ATP to ADP and P_i has been known for some time to be associated with mitochondrial membranes:

$$ATP^{4-} + H_2O \xrightarrow{\text{ATP-ase}} ADP^{3-} + H_2PO_4^{2-} + H^+.$$

Mitchell suggested that oxidative phosphorylation might occur when this ATP-ase reaction was reversed, because of its orientation in the membrane and of the effect of the charge separation taking place simultaneously at the oxidation–reduction step. The mechanism connecting the oxidation–reduction step with the compulsorily linked phosphorylation had long been obscure and different investigators have suggested a number of different hypothetical steps and intermediates. After reviewing the evidence for various intermediates, Slater (1966) concluded that no compound corresponding to any unknown intermediate, phosphorylated or otherwise, has been isolated. Mitchell attempted to eliminate the need for these hypothetical intermediates by the suggestion that the charge separation might act on the ATP-ase. Since the hydrolysis of ATP by mitochondrial ATP-ase is accompanied by the gain of two hydrogen ions in the external medium (Mitchell and Moyle, 1965), and there is evidence that the ATP and ADP are inside the membrane (Pfaff, Klingenberg and Heldt, 1965), the hypothetical reaction is as shown in fig. 5a. Formally the reaction can be written:

$$ATP + H_2O + 2H_R^+ \xrightleftharpoons[\text{synthesis}]{\text{hydrolysis}} ADP + P_i + 2H_L^+,$$

where the subscript L and R refer to the left-hand and right-hand side of the membrane respectively. If the oxidation–

reduction process is proceeding simultaneously (fig. 5b), the concentration of H^+ on the left will be increased by protons and the H^+ concentration on the right will be reduced by the OH^- combining to form water. This decrease in H_R^+ and the increase in H_L^+ will force the reaction towards synthesis, so ATP would be synthesized by a pH shift across the membrane,

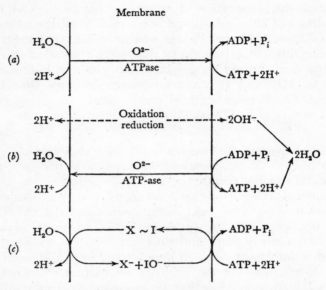

Fig. 5. Principles of Mitchell's hypothesis of the reversible ATP-ase in the membrane of the mitochondrion. (a) Normal ATP-ase with two protons extruded for one ATP hydrolysed; (b) Synthesis of ATP due to protons and hydroxyl ions in simultaneous oxidation–reduction reaction; H^+ concentration increases on the left and due to water formation, decreases on the right, reversing the ATP-ase reaction; (c) Normal ATP-ase, showing the possibility that specific substances X and I are translocating O^{2-} to the right and returning as the corresponding anhydride to the left.

due to the simultaneous oxidation–reduction. In later versions of the hypothesis, Mitchell (1966) recognizes the need to postulate one non-phosphorylated intermediate, $X \sim I$, which is probably an anhydride and is shown hypothetically in fig. 5c. The anhydride breaks down to the hypothetical X^- and IO^- when the oxygen is accepted from the water on the left. Mitchell also recognizes the need to reconcile these hypothetical intermediates with the coupling factors, soluble proteins, which can be isolated from beef heart mitochondria, and which when added back will restore the phosphorylating activity to the respiratory chain of depleted mitochondrial fragments. These factors, originally isolated about the same time by Linnane (1958) and by Pullman, Penefsky and Racker (1958), have been studied in considerable detail by Racker and his collaborators. They are apparently associated with the characteristic knobs which are seen by electron microscopy to be attached to the inner surface of the cristae membranes of mitochondria.

The details of mechanism for the ATP-ase are still not understood but the Mitchell hypothesis requires, in principle, only that it can be reversed by a hydrogen ion differential across the membrane in which it occurs. The evidence for this will be discussed in the next chapter. The essentials are that the separation of charge due to oxidation–reduction can bring about a change in the hydro-dehydration reaction with an ATP-ase which catalyses the ADP, P_i, ATP and water reaction.

Mitchell analyses the energetics of the ATP-ase reaction

$$\text{ATP} + \text{H}_2\text{O} + 2\text{H}_R^+ \rightleftharpoons \text{ADP} + \text{POH} + 2\text{H}_L^+,$$

where the suffixes L and R represent the aqueous phase on the left and right respectively, of the ATP-ase containing membrane.

The normal hydrolysis equilibrium for ATP would be

$$\frac{[\text{ADP}]\,[\text{P}_i]}{[\text{ATP}]} = \text{K}[\text{H}_2\text{O}];$$

but if, as in the above reaction, the hydrolytic reaction is strictly coupled to the movement of two protons from right to left per ATP hydrolysed

$$\frac{[\text{ADP}]\,[\text{P}_i]}{[\text{ATP}]} = \text{K}[\text{H}_2\text{O}]\frac{[\text{H}_R^+]^2}{[\text{H}_L^+]^2},$$

where $[\text{H}^+]$ = electrochemical activity. The assumption is that ADP, P_i and ATP all take part in the equilibrium from the same side of the membrane. If there is a membrane potential of ΔE millivolts between the two sides of the membrane, positive on the left,

$$\log\frac{[\text{H}_L^+]}{[\text{H}_R^+]} = \text{pH}_R - \text{pH}_L + \frac{\Delta E}{Z},$$

where $\mathcal{Z} = 2\cdot303RT/F$, F is the Faraday, R the gas constant and T the absolute temperature.

From the last two equations

$$\log_{10} \frac{[ATP_L]}{[ADP_L]\,[P_{iL}]} = 2\left(pH_R - pH_L + \frac{\Delta E}{\mathcal{Z}}\right) - \log_{10} K[H_2O].$$

When T is $300\,°K$, $\mathcal{Z}$ is close to $60\,mV$ and the hydrolysis constant ($\log_{10} K$) for ATP is close to 5. Substituting ΔpH for $pH_L - pH_R$ and assuming the concentration of P_i to be $0\cdot01\,M$,

$$\log_{10} \frac{[ATP]}{[ADP]} \simeq \frac{\Delta E}{30} - 2\Delta pH - 7.$$

Thus the electrochemical potential difference of hydrogen ion necessary to poise the ATP/ADP ratio at 1 (or as Mitchell says to poise the ATP/ADP couple central), under these conditions would be
$$\Delta E - 60\Delta pH \simeq 210\,mV.$$

A pH differential of $3\cdot5$ units (acid on the left) or a membrane potential of $210\,mV$ or a combination of both would poise the system central under these conditions. It follows that if a potential difference or a pH difference of the appropriate magnitude is imposed across the membrane, the reaction would be shifted towards synthesis of ATP. As we shall see in the next chapter, changes of pH can be shown to occur across chloroplast lamellae and are associated with ATP synthesis.

The Mitchell hypothesis introduced another very important idea. He pointed out that if a hydrogen carrier operates to take hydrogen atoms across a membrane and is acted on by an electron carrier which takes only electrons back across the membrane, protons will be transferred to the side of the membrane opposite to that on which they were picked up (fig. 6). Furthermore, if more than one pair of hydrogen and electron carriers occurred in the chain with the right orientation in the membrane, transport of H^+ ions could occur at each pair. We have referred to the observation that phosphorylation occurs at three sites when substrates passing electrons to the NAD are supplied to the electron transport chain and occurs at two sites when succinate is used as substrate. On the chemi-osmotic hypothesis this would suggest that three sites of hydrogen ion transport are operating from substrate via NAD (fig. 2) and only two sites if succinate is the substrate. Mitchell's suggestion for how the members of the electron-transport chain might be arranged to provide these sites of H^+ transport is shown in fig. 7. Some parts of this hypothetical arrangement of the chain are controversial; for instance, most investigators would put

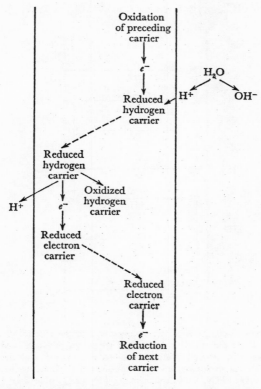

Fig. 6. Mitchell's hypothesis for proton transport from one side of a membrane to the other with a hydrogen carrier in one direction and an electron carrier in the other.

coenzyme Q (ubiquinone) ahead of cytochrome b and many would be uncertain about the place of the non-haem iron and sulphydryl groups. At this stage, the question is whether the hypothesis is useful in principle; the evidence will be discussed in the next two chapters. This aspect of the hypothesis suggests how the passage of electrons through the transport system could be accompanied by transport of protons at three places and could result in three phosphorylations. Further, the same principle would explain those examples where more hydrogen ions and hydroxyl ions are separated (e.g. Davies's ratio for HCl to oxygen of 12, see chapter 1) than only those resulting from the substrate protons and electrons transferred to oxygen (where the HCl/oxygen ratio could not exceed 4).

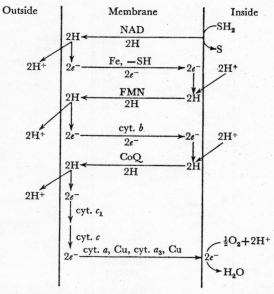

Fig. 7. Mitchell's hypothesis of the arrangement of the electron-transport chain which would allow transport of protons at three places (with NAD-linked substrates) and result in three phosphorylations with each pair of electrons passing to oxygen. NAD, nicotinamide adenine dinucleotide; SH_2, reduced substrate; —SH, sulphydryl; FMN, flavin mononucleotide; CoQ, ubiquinone; cyt, various cytochromes.

The Mitchell hypothesis thus explains a number of phenomena associated with an electron-transport chain, charge separation, phosphorylation, establishment of electric potential differences and possibilities for ion exchange. The basic postulates of the chemi-osmotic hypothesis are (1) a membrane with very low permeability to ions, (2) an oxidation–reduction system located in the membrane with proton–electron conducting properties, (3) an ATP synthesizing and hydrolysing system located in the membrane and controlled by the pH differential on the two sides, and (4) specific exchange diffusion systems which control the ions that can pass from one side of the membrane to the other (to be discussed later).

In the next chapter the evidence for this hypothesis will be given and this will lead to a critical examination of how far the hypothesis describes the principal properties of organelles and, later, of cells.

When this book was written, I had not seen the papers by Williams (1961, 1962) which discuss the separation of proton and hydroxyl in membranes. Williams suggested that the protons in the lipid phase of the membrane were the factor which resulted in the formation of an anhydride of a phenol or of a phosphate. This interpretation has much in common with the Mitchell hypothesis but differs in the details of the mechanism suggested, particularly on the importance of the proton in the membrane itself. Future developments of hypotheses should take account of Williams' suggestions.

CHAPTER 3

Evidence from Organelles—Phosphorylation

'When you can measure what you are speaking about and
express it in numbers you know something about it, but
when you cannot measure it, when you cannot express it in
numbers, your knowledge is of a meagre and unsatisfactory
kind'.   (KELVIN, 1883)

THOUGH Mitchell had developed the chemi-osmotic hypothesis mostly with the electron transport chain of mitochondria
in mind, the most dramatic evidence for the hypothesis came
from work on chloroplasts and, in particular, from that of
Jagendorf and his associates. A very readable account of this
work by Jagendorf and Uribe (1967) is recommended; it has
been much used in the preparation of this chapter.

Isolated chloroplasts

Photosynthetic work on isolated chloroplasts had been going on
for many years. The chloroplast in the green leaf is a larger,
even more complicated structure than the mitochondrion.
Characteristically it consists of a series of membranes, arranged
as flat sheets or plates, and known as lamellae, distributed
through a matrix or stroma which is enclosed in an envelope of
two membranes. In some parts of the plastid the lamellae are
more tightly packed than in others, and in such places, being
circular in surface view, resemble piles of disks and constitute
the grana. Intact chloroplasts can be isolated from the green
leaf and purified free from other cell constituents by differential
centrifuging. Under many conditions of isolation the plastids
lose the outer envelope and the contents of the stroma. Naked
lamellae remain and, though swollen and separated in varying
degrees, carry out many of the reactions of photosynthesis.
Isolated swollen plastids of this type have provided evidence
that the lamellae can bring about the separation of charge
which is the basis of this hypothesis and can undertake ATP

formation as the result of the charge separation. Lamellae are composed mainly of lipoproteins and contain the chlorophyll, carotenoid and cytochrome pigments which are essential to the photosynthetic process.

Details of the photosynthetic process are still being investigated and not all the stages are understood, but much accumulated evidence points to the participation of two light reactions, one of which is concerned with the transfer of electrons and protons from water (fig. 8 a). This reaction, known as light-reaction II, is not activated by light of wavelengths greater than 705 mμ. Absorption of light results in the liberation of oxygen from water and the transfer of protons and electrons to the hydrogen carrier, plastoquinone (Hill and Bendall, 1960). The plastoquinone transfers the electrons to cytochrome, which is thought to pass them on to the chlorophyll concerned with the other light reaction, light-reaction I, which is activated by light of wavelengths up to 730 mμ. This light reaction results in the passage of electrons to ferredoxin, which passes them on to the nicotinamide-adenine dinucleotide phosphate (NADP) where protons are acquired again. The reduced NADP is the reductant for the reactions which turn carbon dioxide into carbohydrate and which, for our present purpose, need not be discussed further. As I pointed out in 1960, the stages associated with the electron and hydrogen ion transfers could be responsible for a physical separation of electrons and hydrogen ions if the electrons passed across a membrane (or lamella) impermeable to hydrogen ions. In fig. 8 b the electron–hydrogen transport is arranged to show how charge separation according to Mitchell's hypothesis could occur. It is generally agreed that somewhere near where electrons pass through the electron carriers and the plastoquinone, ATP is simultaneously synthesized from ADP and P_i—the process of photophosphorylation.

Shen and Shen (1962) and, at about the same time, Hind and Jagendorf (1963), with wheat chloroplasts and broken spinach chloroplasts (naked lamellae) respectively, discovered that the light reaction, without phosphate present, enables the plastid to make ATP from ADP and P_i in a subsequent dark period. The reaction in the light was markedly stimulated by a redox dye (which presumably stimulated electron flow). The presence of a non-phosphorylated 'high-energy intermediate', which was formed by the redox reactions and stored up, was

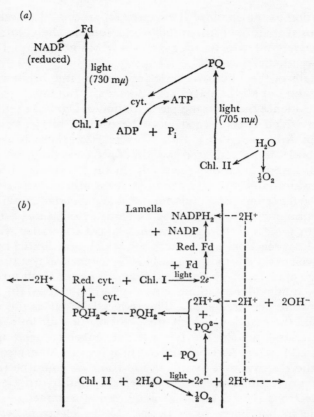

Fig. 8. (a) Simplified scheme of the electron pathways in photosynthesis, set out to represent energy levels with light reactions transferring the electrons against the thermochemical gradient and the hydrogen and electron carriers transferring them 'downhill'. The accompanying proton movements are not shown. Electrons are transferred from water to NADP, and ATP is synthesized. (b) Simplified scheme of electron pathways set out to show how protons might be transferred from right to left by the hydrogen carrier plastoquinone (PQ) which is oxidized on the other side of the lamella by the electron carrier, cytochrome. The net result is that two H^+ would be transferred across the lamella with the passage of each pair of electrons from water to NADP; a corresponding increase in OH^- occurs on the opposite side. Chl, chlorophyll; cyt, cytochrome; fd, ferredoxin. Solid arrows, chemical reactions; broken arrows, diffusion pathways.

inferred. This apparent 'intermediate' subsequently caused ATP formation. The formation of the intermediate (if it existed) was complete in 30 s of illumination when pyocyanine was the co-factor. The loss of the 'intermediate' in the dark, as its yield in the light, varied with pH. An inhibitor of electron flow such as dichlorophenyl dimethylurea (DCMU) was inhibitory when added in light but had absolutely no effect when added in the dark. Arsenate, on the other hand, which might be competitive with phosphate at the phosphorylation stage, inhibited only in the dark. Every one of a large range of uncoupling reagents accelerated the dark decay process very considerably, and some provided an unusual pattern of events —such as a greater inhibition when added in the dark than if present during the initial light period (atebrin) or a failure to inhibit the yield even though dark decay was accelerated (chlorpromazin). One of the puzzles about the evidence for an intermediate as an energy-rich compound was its abundance. It was possible to make about 270 molecules of ATP per set of 2500 chlorophyll molecules. This large number ruled out an intermediate as a stoichiometric compound involving any of the usual electron transporting agents, though it probably represents only about half the potential ATP synthesis, the other half being lost under the experimental conditions.

The probabilities of an 'energy-rich intermediate' as the result of the electron carrier complex seemed low. Jagendorf and Uribe are interesting to quote, showing the way scientific thought often advances. 'At the time when our confidence in a high energy state of an electron carrier was waning, Dr Hind spent some profitable time in the library and there discovered the chemi-osmotic hypothesis presented in the 1961 paper by Mitchell...we decided to consider seriously at least for the moment, the possibility that our energetic intermediate was really a pH gradient. One immediate prediction from the theory was that the pH of the external medium ought to increase as [the intermediate] forms in the light, if the protons are moved into the internal space of the grana disk sacs and become inaccessible to the glass electrode. This rise in pH, caused by illuminating a suspension of unbuffered or weakly buffered chloroplasts, was indeed found to occur (Jagendorf and Hind, 1963; Neumann and Jagendorf, 1964). With this discovery our opinion as to the probability of the chemi-osmotic hypothesis was revised upwards several notches, and the nature of our experiments has never been the same since.' The concluding part of that quotation emphasizes how, as Jagendorf and Uribe point out earlier in the article, the nature of a guiding hypothesis in science can have a considerable effect on the kind of experiments one might wish to attempt.

It was then clear that the search for a quantity of high-energy intermediate was nearing an end. The pH rise (as much as 1·0 units) had approximately the same pH and co-factor requirements as the 'intermediate' which had been sought. The rise time for the pH change was about 30 s and the decay somewhat longer. With pyocyanine as co-factor the rise and decay were faster and the yield greater than with other co-factors or with none. Phosphorylation uncouplers all depressed the extent of the pH change and caused a faster decay in the dark after illumination. Further, non-specific detergents which dissolve the membranes destroyed the pH differential between the inside and outside of the grana sacs. On the chemi-osmotic hypothesis uncouplers would be considered as making the membranes leaky to protons. Altogether, the suggestion of a hydrogen ion change resulting from the light reaction looked more reasonable than the build-up of a hypothetical high-energy 'intermediate' which could be so readily destroyed by uncouplers and detergents. Further, the amounts of acid formed internally—which could be measured by adding acid to compensate the increase in alkalinity in the external medium—seemed to argue for ion movement, rather than for a specific chemical reaction. Often about 1800 protons would be transported per 2500 chlorophyll molecules and sometimes more—up to about one proton per chlorophyll molecule. No single catalytic component of the plastid is so abundant. The calculated internal pH, if all the hydrogen ions were free in solution and reasonable assumptions were made about the volumes between the lamellae, would have been about 2·5. If buffering groups inside could neutralize 90% of the protons entering, the pH would still be as low as 3·5. As the experiments for synthesizing ATP in the dark required the medium pH to be 8, the difference between inside and out would have been 4·5 pH units, which would be more than enough to drive the reversible ATP-ase of Mitchell's hypothesis to synthesis of ATP. One of the principal difficulties in reconciling the chemi-osmotic hypothesis with the facts of experiment was the observation that the pH continued to rise after the light was switched off, following brief periods of illumination. Izawa and Hind (1967) subsequently showed that this was due to a technical difficulty in measuring the pH shift without a lag in time and that, using improved techniques, the pH rise in the external medium was seen to stop abruptly after the light was turned off. Cation excretion from the chloroplasts accompanied the build-up of internal hydrogen ions and we shall return to this phenomenon later in the book. For our present purposes we can agree with Jagendorf and Uribe that since the kind of cation excreted was variable and the amount of hydrogen ions taken up was greater than any one cation, it is likely that the hydrogen ion uptake preceded the cation excretion.

All this evidence pointed to the possibility that a gradient in hydrogen ion activity, not an energy-rich intermediate, was responsible for the ATP synthesis. This hypothesis could be tested by the simple experiment of soaking some chloroplasts in acid in the dark to allow some hydrogens ions to leak in and then transferring them, still in the dark, to the alkali reaction medium at pH 8 containing ADP, phosphate and magnesium. Dr Hind discovered that this resulted in the synthesis of ATP and later work showed that use of an appropriate organic acid as a buffer would increase the yields (Hind and Jagendorf, 1965, Jagendorf and Uribe, 1966). With these techniques a total number of 550 ATP molecules per 2500 chlorophyll molecules could be synthesized. This amount could not result from any simple stoichiometric relationship to the quantity of any electron carrier. The ordinary uncouplers of photophosphorylation and the energy-transfer inhibitors inhibited ATP formation by this system. Perhaps most important, the antibody to the chloroplast terminal enzyme of phosphorylation, which inhibits both the latent ATP-ase of the enzyme and the photophosphorylation, also inhibited the acid-to-alkali phosphorylation (McCarty and Racker, 1966a, b). Thus the ordinary terminal enzymes of the photophosphorylation process are concerned. Two pieces of evidence argue strongly for the reaction being due to the pH differential—the hydrogen ions themselves: first, the nature of the accompanying anion was unimportant except in as far as it governed the pH differential; secondly, if the particles were allowed to swell so that the volumes in the grana sacs were increased and greater amounts of organic acid could be included, greater amounts of ATP were synthesized.

There are many other interesting aspects of this work on isolated chloroplasts, some of which will be referred to elsewhere. The principal observations have been confirmed by others. Furthermore, Mayne and Clayton (1966) have shown that a pH shift between inside and outside the lamellae could be used to make them luminescent, so that energy due to separation of charge can reverse the reactions right back to the photochemical step at chlorophyll. Summarizing this discussion we may conclude: (1) that a light reaction brings about a separation of H^+ and OH^- on opposite sides of the chloroplast lamellae; (2) that a pH differential with hydrogen ions on the inside of the lamellar sacs and hydroxyl ions on the outside will result in the synthesis of ATP by an enzyme which shares many of the properties of the photophosphorylating enzyme; (3) that photophosphorylation is the result of the charge separation (consequent on the absorption of light) acting on this enzyme system—strong evidence for the validity of the chemi-osmotic hypothesis for chloroplast lamellae.

Isolated mitochondria

The demonstration that similar processes, separation of positive and negative charge, and pH differential leading to phosphorylation, take place in mitochondria has proved more difficult. This might be expected because the volumes of mitochondria are much less than those of chloroplasts, so the reservoir of ions inside the individual mitochondrion would be less, particularly as the whole internal volume might not be available. Significant progress with this system has been made, particularly in experiments by Mitchell and Moyle, which will be discussed in detail. The experiments are more difficult to do and the interpretation less unequivocal than those on chloroplasts. Nevertheless there is good evidence that mitochondrial phosphorylation shows some characteristics similar to that of chloroplast phosphorylation and consistent with the chemi-osmotic hypothesis—though mitochondria differ from chloroplasts in producing hydrogen ions on the outside of their membranes and hydroxyl ions on the inside.

Following the initial discovery by Vasington and Murphy (1961), there has been renewed interest in the problems of ion uptake by mitochondria in recent years. Mitochondria taking up oxygen will extrude protons under certain conditions (Brierley, Bachmann and Green, 1962; Chappell and Greville, 1963). These experiments were done under conditions where a certain amount of uncoupling of phosphorylation had taken place and could best be demonstrated by using divalent phosphate salts. In the absence of ADP or in the presence of ADP with oligomycin (so ATP formation is prevented), increased oxygen uptake was accompanied by massive accumulation of divalent phosphates precipitated inside the mitochondrion, and by extrusion of hydrogen ions. Those of us who thought in terms of proton release by the electron transport systems of the mitochondrial membrane immediately saw this extrusion of hydrogen ions as an example of this phenomenon and interpreted the results in this way (Millard, Wiskich and Robertson, 1964; Chappell and Crofts, 1965). To others, this extrusion of hydrogen ions which is accompanied by an uptake of cations was seen as being mainly due to exchange for hydrogen ions already present in the mitochondria. It is not proposed to discuss the details of this controversy but rather to outline the experimental results, which seem to prove that, at least under some conditions, the protons of the electron transport chain must be the principal source of the hydrogen ions extruded. Snoswell (1966) investigated the constant increase in hydrogen ion concentration of the medium in which rat liver mitochondria were respiring in a medium containing $0 \cdot 17 \text{M}$ sucrose and 40mM choline chloride. Proton extrusion continued for 1 h or more, if respiration continued, and was not sensitive to oligomycin, though it was stopped immediately by antimycin. The amount extruded was too great for it to have been derived from any stored ATP. Also the lack of inhibition by oligomycin makes ATP breakdown very unlikely as a source of the hydrogen ions. Again, if the protons were coming from ATP, it is not likely that their extrusion would be stopped by a respiratory inhibitor such as antimycin. Cations from the external medium, exchanging for hydrogen ions already in the mitochondrion, will not explain the continued extrusion in Snoswell's experiments since there was no obvious source of cations in the medium for exchange reactions, and the release of hydrogen ions was observed when the mitochondria were in $0 \cdot 25 \text{M}$ sucrose alone, although the rates were unreliable in that medium. Furthermore, in these experiments, addition of ions such as Mg^{2+}, $H_2PO_4^-$ and K^+ had little or no effect on the rate of proton extrusion. The addition of antimycin stopped the decrease in external pH but was not followed by an increase as

27

might have been expected if the hydrogen ions could exchange for cations which had been taken up when they were extruded.

Snoswell found that the rate of proton extrusion was related to the rate of respiration but also varied with the substrate supplied. The highest ratio of proton extrusion to oxygen uptake was obtained when endogenous substrate was being respired. For example, the rate of extrusion was 25 mμ moles H$^+$/min/mg mitochondrial protein with endogenous substrate giving an oxygen uptake of 21 mμ atoms/min/mg mitochondrial protein. If we assume, according to the original (Robertson, 1960), hypothesis, that two protons might be extruded when two electrons are transferred to oxygen, this could be interpreted as a charge separation with about 60% efficiency. Any tendency for the membrane to leak ions or to short-circuit charge would decrease the efficiency of the charge separation. This probably accounts for Snoswell's observations that the addition of substrate ions decreases the H$^+$/O ratio though they may increase the oxygen uptake (succinate) or maintain it at the same rate as the endogenous substrate (glutamate or pyruvate plus malate). Snoswell's experiments with dinitrophenol, added as an uncoupler in the presence of pyruvate and malate, showed that oxygen uptake might be increased from 21 to 75 mμg atoms O/min/mg mitochondrial protein and the hydrogen ion extrusion rate simultaneously was increased from 11 to 25 mμmoles H$^+$/min/mg mitochondrial protein. The 2H$^+$/O ratio or the efficiency of proton extrusion per electron transfer dropped from about 27% to about 17% in this experiment. The simplest interpretation is that the dinitrophenol as an uncoupler enhances the separation of protons and electrons but it also increases the leakiness of the membrane so the separation is not maintained so efficiently. This aspect of uncouplers will be discussed subsequently.

One other result of Snoswell's work is interesting as a model. He used ferricyanide in the presence of cyanide as an electron acceptor instead of oxygen: electrons are accepted predominantly at the cytochrome-c level. In a medium containing trischloride, succinate and 10^{-3}M cyanide and no oxygen there was no continued change in the external pH. The addition of 1 mM ferricyanide caused a very rapid liberation of H$^+$ ions, equal in amount to the ferricyanide; that is, protons remained in the solution when the solute was reduced by the electrons.

By contrast, if oxygen were the electron acceptor instead of ferricyanide, the electrons, oxygen and water would result in the formation of hydroxyl ions. If hydrogen and hydroxyl ions are completely separated by the mitochondrial membrane as suggested by the Mitchell hypothesis, the rate of proton extrusion would be equivalent to the rate of electron transport. Any less rate of proton extrusion would be a measure of the permeability of the membrane to hydrogen or hydroxyl ions which coming together would form water. In Snoswell's experiments proton extrusion was always highest in freshly prepared mitochondria and decreased as they were aged at 0° C until, after 24 h, no respiratory-dependent proton extrusion occurred.

The most detailed quantitative measurements of the hydrogen-ion extrusion have been done by Mitchell and Moyle (1965, 1967). They took a suspension of rat liver mitochondria in potassium chloride, glycyl-glycine buffer at pH 7·2 and substrate, either β-hydroxybutyrate or succinate, and incubated at 25 °C in the absence of oxygen. When small amounts of oxygen were introduced there was an absorption of oxygen and a rapid production of hydrogen ions in the medium. Subsequently the hydrogen ion in the external medium decayed slowly. With β-hydroxybutyrate as substrate, sixteen measurements of the H^+/O ratio on four different batches of mitochondria gave a mean value of 5·90 ± 0·33. Succinate used as substrate in the presence of rotenone (which stops oxidation linked to NADH) gave a mean value of 3·25 ± 0·08 at 25 °C and a mean value of 4·07 ± 0·11 at 5 °C. These quantitative data are in agreement with the expectation on the chemi-osmotic hypothesis that if an NADH-dependent substrate like β-hydroxybutyrate is used, there would be three places at which protons and electrons would be separated across the membrane, and if succinate is used there would be only two. These results were confirmed in later work by Mitchell and Moyle (1967), investigating different spans of the electron-transport chain and the effects of inhibitors and uncouplers. The measurements were consistent with the Mitchell hypothesis that three separate sites of proton translocation across the mitochondrial membrane are associated with the electron-transport chain; uncouplers are believed to bring about a flow of protons back through the membrane, so the hydrogen and hydroxyl ions do not remain separated long enough to give the build-up of external acid.

Demonstration that the mitochondrion can make ATP when acted on by a pH gradient has proved difficult. A number of workers who have tried to do this have failed, but Reid, Moyle and Mitchell (1966) obtained the synthesis of a small amount of ATP when the pH of the medium was lowered from about 9 to 4·3. This experiment was carried out in the presence of K^+ and of valinomycin, which increases the permeability of the membrane to K^+. The amount of ATP formed was greater than that expected from the effect of the hydroxyl ion on the ADP, P_i and ATP equilibrium. D. L. Millard and J. T. Wiskich (personal communication) have obtained similar results with isolated beetroot mitochondria. The mitochondria were treated with alkali at pH 10 for 1 min and then with acid at about pH 6 in the presence of ADP, glucose, hexokinase and phosphate. Antimycin A was added to inhibit endogenous substrate oxidation and AMP to inhibit adenylate kinase. Significant but small amounts of ATP were formed in this base-to-acid transition treatment and none was formed when the pH was not varied. ATP formation was inhibited both by oligomycin and by DNP.

Mechanism of phosphorylation

The various difficulties in the chemi-osmotic hypothesis applied to mitochondria have been set out by Slater (1967), who reviews the earlier criticisms, particularly those of Chance and Mela (1966 a, b) and Tager, Veldsema-Currie and Slater (1966).

Though Slater and Chance have doubts about the validity of the Mitchell hypothesis as to the primary act, Slater suggests that the data may be interpreted according to two other schemes, both of which also put the H^+ extrusion close to the phosphorylation. These schemes may be compared in fig. 9, where E_c is an energy-rich intermediate and C^{n+} is the cation involved.

The second alternative is unlikely for mitochondria, as we have seen from Snoswell's results, but it is ruled out by the evidence for isolated chloroplasts, where the primary act seems to be the proton separation and there is insufficient exchange of cation to account for the hydrogen ions accumulating in the grana sacs. Further, the evidence for the nature of the ATP-forming enzyme, summarized by Jagendorf and Uribe, suggests that it is not necessary to postulate any intermediate between the hydrogen ions and the phosphorylation and thus reduces the probability that the first alternative is correct. The Mitchell hypothesis recognizes the non-phosphorylated intermediate, which is similar to Racker's coupling factor but is associated closely with the reversible ATP-ase and is not an intermediate of oxidation reduction.

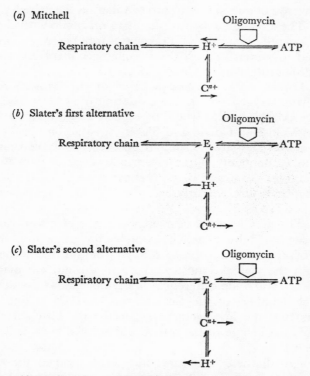

Fig. 9. Alternative schemes for connecting phosphorylation, hydrogen ion extrusion and cation uptake (as suggested by Slater, 1967).

The difficulties in accepting the Mitchell hypothesis, as summarized by Slater, with my comments in brackets, are as follows:

(1) Mitochondria which are phosphorylating cannot be shown by inhibitor experiments to be extruding protons (Chance and Mela, 1966a, b). (This would not be expected, so

31

the criticism is not valid. A small and temporary build-up of H^+ concentration would be very difficult to demonstrate.)

(2) Though the reversible ATP-ase has been demonstrated to occur (Reid *et al.* 1966; D. L. Millard and J. T. Wiskich, personal communication), this is insufficient evidence to postulate that the proton gradient is the primary energy conserving act; cations are in general required for extrusion of protons. (It would be surprising if the clear-cut results with chloroplast fragments were not due to the proton gradient as the primary step.)

(3) The stoichiometry is difficult to reconcile with the process (Cockrell, Harris and Pressman, 1966). (More work is required to understand all the ion changes involved; some of these difficulties will be discussed later.)

(4) Under the conditions of Mitchell and Moyle's experiments with oxygen pulses, the oxidation of NADH associated with the extrusion of H^+ would be expected and this is not found (Tager, Veldsema-Currie and Slater, 1966). (The problems of steady-state reactions, dependent on the velocity constants, of different components of the electron transport chain are still considerable. We would need to know more about the possible significance of this difficulty before it could be used to rule out the hypothesis.)

(5) The arrangement of members of the electron-transport chain suggested by Mitchell to allow for three hypothetical points of separation of protons and electrons is difficult to reconcile with the views of most workers on the current evidence. (There are still wide differences of opinion about the arrangement of members of the electron-transport chain and many workers find the arrangement suggested by Mitchell no less probable than some of the others which have been tentatively accepted.)

Though all these criticisms must be examined thoroughly with further experiments, none alone or combined with others seems to be sufficiently definitive to rule out the hypothesis. We may conclude that quantitative evidence suggests strongly that the separation of proton and electron, leading to separation of hydrogen ions and hydroxyl ions across a membrane, is a fundamental process in both plastids and mitochondria and is probably the basis of the phosphorylation mechanism. In the next chapter we shall discuss how this is related to ion movements across the membrane.

CHAPTER 4

Charge Separation, Ion Movement, Swelling and Shrinking in Organelles

Did the chicken or the egg come first?

ORGANELLES, such as chloroplasts and mitochondria, would be expected to have characteristic but changing ionic composition. First, such bodies would contain a number of metabolic anions (nucleic acids, proteins, anionic intermediates of metabolic reactions) which would be indiffusible and would therefore attract mobile cations. Secondly, these organelles have membranes, both on the outside and inside, which are relatively impermeable to ions. Thirdly, metabolic activities will change the ionic composition and result in ionic fluxes between the inside and the outside of the organelle. Some of these fluxes will be passive; others may be active due to metabolic pumps. This ion balance, itself an integral part of the metabolic function of the organelle, is of great biological interest and merits investigation. In addition, the balance of ions between the organelle and the surrounding medium will influence the osmotic pressure difference and hence control the swelling and shrinking behaviour. Despite the importance of these problems, little investigation of chloroplasts was carried out until recently. Investigation of electrolyte balance in mitochondria occurred in two bursts—the first in the early 1950s and the second, more intense, in the last four years.

Principles of ion movement

The thesis of this book is that the charge separation across living membranes provides the unifying hypothesis for the explanation of these phenomena. It is desirable to note the ways in which charge separation can bring about the movement of ions.

First, mobile ions will adsorb at a surface such as a membrane or a lamella if a fixed charge of sign opposite to the mobile ion is generated in the surface. For example, a reduced electron carrier fixed in a membrane could carry a negative charge and, if the proton had been separated to the other side of the membrane, a mobile cation would be attracted. Such a cation would remain in the vicinity of the membrane and would not be capable of diffusion (except with another cation to exchange) into the bulk aqueous phase of a mitochondrion or chloroplast (fig. 10 a). Under experimental conditions of swollen lamellae, for instance, any quantity of ion of one charge in the aqueous phase must be accompanied by a similar quantity of ion of opposite charge.

Secondly, a proton may be transferred by a hydrogen carrier from the inside of a membrane to the outside, leaving a metabolic anion on the inside of the membrane in the aqueous phase (for example, an intermediate of the Krebs cycle in a mitochondrion). If the membrane is permeable to a cation, entry of the cation can be expected to balance the metabolic anion. Thus translocation of a proton through the membrane not normally permeable to hydrogen ions leaves a mobile anion, but trapped behind a membrane (fig. 10 b).

Thirdly, proton transport may be coupled with anion transport. The precise mechanism by which this occurs may be of prime importance in this hypothesis and will be examined further below. After the movement of H^+ with A^-, a cation could exchange for the H^+. Both cation and anion would then accumulate inside, on one side of the membrane, though only water (formed from the proton and hydroxyl) would appear on the other (fig. 10 c).

Finally, changes in hydrogen ions will affect the ionization of any weak electrolytes present. As the inside of a chloroplast, for example, becomes acid, the ionization of weak anions would be suppressed and, as the membranes are more permeable to molecules than to ions, the undissociated acids would diffuse out (fig. 10 d).

(a) *Coupled anion movement*

We are still lacking conclusive experimental evidence about the first ionic movements which accompany the charge separation of H^+ and OH^-. We have seen that the H^+ could not move from the membrane in which it was separated from the electron

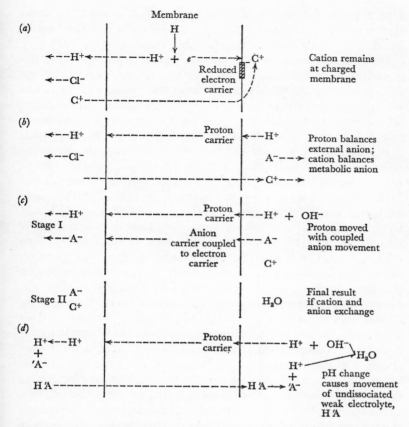

Fig. 10. Principles of ion movement and charge separation. (a) When a negative charge remains in the membrane and a mobile cation (C^+) is adsorbed. (b) When proton transport occurs, leaving an anion behind the membrane to be balanced by an entering cation. (c) When proton transport is accompanied by anion transport (stage I: acid is extruded on the left and alkali accumulates on the right, the cation balancing the hydroxyl; stage II: membrane allows cation to exchange with H^+ and water is formed). (d) When the acidity due to proton transport decreases the ionization of a weak electrolyte ($H'A$) and the membrane is permeable to the undissociated acid. The net result is movement of anion from left to right. This would tend to be accompanied by a corresponding movement of cation through a cation permeable area.

unless it were accompanied by a mobile anion. Similarly, the OH^- formed in oxygen uptake by the electron transport chain or by the splitting of water in photosynthesis could not move from the membrane unless accompanied by a mobile cation.

35

Unless the hydrogen ion and the hydroxyl do move away from the membrane the net charge across the membrane would build up sufficiently to oppose the movement of reducing equivalents through the electron transport chain and the oxygen uptake would stop. This may be the explanation of the situation observed in mitochondria in the absence of an anion which will penetrate (for example, phosphate or acetate, Chappell and Crofts, 1966) when the cation-induced oxidation rate is inhibited (the condition referred to by Chance, 1964, 1965, as state 6). Since the divalent cation uptake in mitochondria can be independent of the phosphate uptake (Chappell, Cohn and Greville, 1963; Chance, 1965) and since in the presence of valinomycin or gramicidin, alkali metal ions exchange for H^+ ions (Pressman, 1963, 1965; Chappell and Crofts, 1965, 1966), it has been assumed that the H^+ extrusion and cation uptake are more basic than anion uptake. This evidence does not exclude the possibility that the proton–electron separation was accompanied by anion movement in the first place. Lundegårdh (1945) first suggested that the electron transfers in the cytochrome system were compulsorily linked to an anion transport. This became the basis of his hypothesis of anion respiration in plants and has been his view ever since; Lundegårdh (1966) has recently restated this hypothesis. On theoretical grounds he believes that the cytochrome system cannot conduct unit negative charge as an electron to one side of a membrane without carrying unit negative charge as an anion to the other. In support of this hypothesis he quotes the observations of Boeri and Tosi (1954) that the *in vitro* oxidation of reduced cytochrome *c* is accelerated by the presence of free salt anions. Lundegårdh (1953) had shown that the oxidation of reduced cytochrome *c* by a homogenate containing mitochondria was accelerated by ions. Similar results were obtained by Miller and Evans (1956) and by Honda, Robertson and Gregory (1958), who also showed that ions accelerated the rate of NADH oxidase but these results are complicated by the possibility of ionic strength effects on mitochondrial structure. The hypothesis that an anion is transported when an electron goes in the opposite direction is also necessary to explain the chloride transport in gastric secretion; as pointed out in chapter 2 there is good evidence that the chloride is actively transported in gastric HCl secretion, though the hydrogen ions are derived simultaneously from the charge separation. Further discussion in this chapter will be based on the hypothesis originally suggested by Lundegårdh (1945) and stated in different form by Robertson and Wilkins (1948 *b*):

The primary step in charge separation by an electron carrier is transport of an electron away from a proton, compulsorily coupled to the transport of an anion to the proton (see fig. 4).

If the charge separation is maintained (by the pH difference across the membrane of a mitochondrion or a plastid), the hydrogen ion can leave the membrane accompanied by the mobile ion. If there is no mobile anion (sometimes called the permeant anion in the mitochondria) on the correct side of the membrane, the electron transport will slow down and the respiration and other consequent changes will stop. Anions must always be present to stimulate changes in mitochondria, and, as the H^+ ions are extruded, this hypothesis requires that anions are transported out of the mitochondrion in proportion to the H^+ ions. Entry of anions can at times limit the rate of oxygen uptake and a permeant anion, as we shall see when considering swelling and shrinking of mitochondria, can be used to initiate oxygen uptake and the processes associated with it (Packer, Utsumi and Mustafa, 1966).

(*b*) Exchange diffusion

The characteristic feature of the organelle membranes is that they are impermeable to ions, including the H^+ and OH^- ions. Consequently the pH gradients established across these membranes are not lost quickly (unless the ATP-ase is operating). If ion transport takes place, the cation can exchange for the hydrogen ion or the anion can exchange for the hydroxyl ion. The second characteristic of these membranes is that the exchange diffusion has a degree of specificity, that is, not every anion can enter the membrane in exchange for the hydroxyl or not every cation can enter in exchange for the hydrogen. With cations, for instance, calcium will penetrate the membrane readily whereas the smaller monovalent potassium will penetrate only very slowly unless the membrane has been influenced by valinomycin or gramicidin. Similarly mitochondria investigated by Chappel and Crofts (1966) are more readily penetrated by the di- and tri-carboxylic acids, glutamate, aspartate, phosphate, ADP and ATP than by the relatively small chloride

37

ion. Given that the initial charge separation has taken place, the potential for charge or ion exchange can be used only if an appropriate ion is available.

Chloroplast ion transport

Despite the early interest in photosynthesis and the recognition that light in green cells increased the rate of uptake of ions, until recently little attention was given to problems of ion balance in chloroplasts. Isolated chloroplasts are recognized as osmometers (Mercer, Hodge, Hope and McLean, 1955) and further as having a preponderance of fixed negative charges. Various workers had shown that the shape and volume of chloroplasts can be altered by light and it was reasoned (Packer, 1963; Jagendorf and Hind, 1963), largely from analogy with

TABLE 1. *Comparison of light-induced structural changes in spinach chloroplasts*

	546 mμ scattering (%)		Volume (μ^3)		546 mμ absorbance	
	Dark	Light	Dark	Light	Dark	Light
Sodium chloride 0·1 M	100	160	45	80	0·8	0·4
Sodium acetate 0·1 M	100	260	45	15	0·8	1·0

mitochondria, that chloroplasts might possess an active mechanism for accumulating ions. Some of the work has led to confusion because the action of light in different circumstances sometimes causes a swelling which is not fully reversible and is therefore damaging to the organelle. Most of this work has been done with grana and not with the intact chloroplast. The light-induced shrinkage is most readily observed with chloroplast fragments in media containing the salts of certain organic acids and in conditions favourable for light-triggered ATP hydrolysis (Packer and Marchant, 1964). The light-induced swelling is observed in strong electrolytes such as sodium chloride and in certain amine salts (Packer, Siegenthaler and Nobel, 1965; Izawa and Good, 1966). The comparative effects are illustrated by table 1, taken from Packer, Deamer and Crofts (1967), in which the light scattering, Coulter counter volume and absorbance were observed. Light scattering by the chloroplast suspension on illumination increased in both solutions, but much more in the acetate than in the chloride; this suggests light-induced structural changes in both systems. The Coulter counter volume changes and absorbance changes indicate that, in the presence of NaCl, swelling occurs since the volume increases and the absorbance decreases, but in the presence of sodium acetate where the opposite effects are seen, the chloroplasts apparently shrink to a third of their initial volume. These workers also investigated the appearance of the chloroplast fragments with the electron microscope and their permeability to various substances in the dark. The main conclusions of their work can be summarized as follows:

(1) Chloroplasts suspended in sodium chloride undergo structural modifications when illuminated by red actinic light in the presence of electron carriers.

(2) Simultaneously with these modifications (swelling between the grana lamellae) protons enter the chloroplast in a light-dependent process. They postulate that this decrease in internal pH causes many of the changes observed in chloroplasts during illumination. In support of this (a) kinetics of proton transport and light scattering increments are similar in onset and decay; (b) pH activation curves are similar for both processes with optima between 5·6 and 6·5; (c) lowering the external pH in the dark to levels which reproduce the theoretical internal hydrogen ion concentrations causes alterations in light scattering, volume and ultrastructure similar to those induced by light.

(3) By contrast, chloroplasts in weak acid anions show different ultrastructural changes associated with shrinkage. It is suggested that the internal drop in pH due to light-induced hydrogen ion uptake causes the weak anions to associate and these undissociated acid molecules diffuse out through the grana membranes, altering the osmotic pressure and hence the volume. The kinetics and pH dependence of the light-induced scattering and hydrogen ion changes are consistent with this mechanism.

The unified mechanism proposed by Packer, Deamer and Crofts is illustrated in fig. 11, which is taken from their paper. It is noted that they have put a question mark against the light-induced uptake of chloride by the plastid membrane. This light-induced chloride uptake would be expected if, as sug-

D   39

gested in the beginning of this chapter, the essential accompaniment of a proton extruded to one side of a membrane is an anion transported to the same side simultaneously. Basically, the swelling occurring in a strong electrolyte would be due to the uptake of anion, accompanied by the movement of proton, by the appropriate redox system in the chloroplast membrane. At present, apart from the suggestions that this is closely associated with the light reactions in the intact *Nitella* cell (MacRobbie, 1965), we do not know how this occurs but it

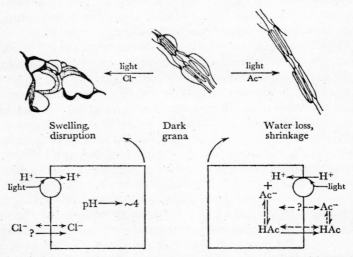

Fig. 11. The unified mechanism proposed by Packer, Deamer and Crofts (1967) for the effect of light on chloroplast ultrastructure.

might well be associated with one of the cytochromes in the membrane. The resulting swelling would possibly be due to two causes: (1) the effect of the decrease in pH on the weak acids of the membrane's solid structure irreversibly altering its properties; (2) the osmotic effect of the ions.

If the anion transported is that of a weak acid, the increasing pH simply prevents the build up of any amount of osmotically active ions because the undissociated acid diffuses out. Hence the lamellar membranes come closer together.

Two other kinds of ion movements in chloroplast fragments must be considered: (1) the replacement of the cations already in the fragment when the light-induced proton uptake starts

(Dilley and Vernon, 1965; Dilley, 1967); and (2) the uptake of divalent phosphate salts first demonstrated by Nobel and Packer (1964a, b).

Dilley and Vernon (1965) investigated the relation between chloroplast shrinkage and ion changes. They found that the volume decrease in chloroplasts, in the presence of light and an electron acceptor, was accompanied by an efflux of magnesium and potassium as well as an increase in internal H^+. All the reactions had kinetics similar to those of the volume change measured by light scattering. Light-induced water, Mg^{2+}, K^+ and H^+ fluxes followed the same pattern of reversal in the following dark period. They suggested that the attachment of protons to negative sites within the chloroplast was the primary event following light. The magnesium and potassium ions and, in some cases, sodium ions thus replaced moved to the exterior of the chloroplast to maintain electric neutrality. A water efflux then occurred to re-establish an osmotic equilibrium. These cation fluxes occurred under conditions suggesting that they may be alternatives to the ATP formation by chloroplasts. More recently Dilley (1967) has extended this model to include the possibility that there is a compartment of mobile cations associated with immobile anions. This compartment is considered to shrink and swell in response to changes in external pH, provided that the H^+ ions could exchange for the internal cations. It is also necessary to postulate that the compartment is bounded by a semi-permeable membrane and is not merely a system of colloid particles with fixed negative charges. Further, it became apparent that the behaviour of the chloroplast as a whole, with its outer membrane intact, was different from the behaviour of the chloroplast fragments or swollen grana. Summarizing the properties of the intact chloroplast, Dilley suggests that there is:

(1) An outer membrane, which is relatively permeable to sucrose and salts, but not to albumen.

(2) A hydrogen–cation exchange mechanism which can be affected either by changing the outside pH or by coupling to the light-induced electron flow. This mechanism functions at the grana membranes.

(3) An osmotic compartment between the grana, which can function as a true semi-permeable membrane to solutes such as sucrose and salts, that is, it gains or loses water in response to

41

changing external tonicity. It can also act as a Donnan system of fixed negative charge surrounded by a membrane which allows the cations to pass only via a hydrogen ion/cation exchange mechanism.

Anions are considered as permeating the membrane freely. It is suggested that this allows changes in the concentration of internal salts and the size of the compartment, due to degree of ionization of fixed charge groups in the compartment, either by changes of external pH or by activation of light-induced hydrogen ion/cation exchange reactions. It will be noted that Dilley's suggestion places the emphasis on the protons known to accumulate inside the granum spaces being balanced by the fixed negative charges in the membrane. As pointed out in the introduction to this chapter, this could be true, but would apply only if the hydrogen ions were maintained in the immediate vicinity of the fixed negative charges. It also neglects what I believe to be the basic system whereby an anion is transported in to accompany the proton. For this reason, among others, the model of Packer, Deamer and Crofts (1967) seems more reasonable.

Nobel and Packer (1964 a, b) first showed that spinach chloroplasts in $vitro$ have a light-dependent uptake of calcium (^{45}Ca) and phosphate (^{32}P). K^+, Rb^+, Mn^{2+}, Zn^{2+}, Fe^{2+}, SO_4^{2-}, I^-, Br^- and Cl^- were apparently not accumulated but strontium could replace calcium in the phosphate-dependent accumulation. Sodium was also taken up but was not phosphate-dependent. The uptake of divalent phosphates seemed to depend on light-activated ATP hydrolysis and the optimum conditions for the adenosine triphosphatase were light, pH 8, Mg^{2+}, the presence of phenazine methosulphate and a thiol compound. ADP and NH_4Cl inhibited the reaction of this ATP-ase. As considerable enhancement of the ATP-ase activity was obtained in the dark following a light pre-incubation, the action of light was considered to be indirect; ion uptake also occurred in the dark after light pre-incubation but both ATP hydrolysis and ion uptake decay in the dark. Since the adenosine triphosphatase, ion uptake and light-scattering changes (indicating changes in chloroplast volume) are all influenced by the same conditions, it was concluded that ion translocations, structural changes and water movements might all be controlled by the same process.

42

Subsequently Nobel, Murakami and Takamiya (1966) and Nobel and Murakami (1967) presented electron-microscope evidence of electron-dense particles on the lamellae of swollen chloroplasts incubated in the presence of strontium, calcium or barium. It was suggested that the electron-dense particles were probably the phosphates of the divalent cations and their number was greatly increased by light under conditions favouring the hydrolysis of ATP. The swollen chloroplasts were mostly without an outer membrane. As chloroplasts swell in light the lamellae progress from the layered state of opposed membranes, characteristic of the intact chloroplast with grana, to a vesicle-like structure, approximately circular in cross-section. The electron-opaque particles appear to be situated on the inner surface of the lamellae.

Recently Nobel (1967) has attempted to distinguish whether the ATP-ase associated with the uptake of calcium in both light and dark is responsible for the transport of calcium or merely functions by providing inorganic phosphate in a convenient location for the precipitate to form. Light greatly increased the uptake of calcium and the uptake of ^{32}P from the γ-position in ATP but had no effect on the uptake of either inorganic phosphate or ^{14}C-labelled ADP under the same conditions. In the absence of calcium no uptake of inorganic phosphate or of ATP was observed. Most of the phosphate accumulated came from the ATP and not from the inorganic phosphate. These results would be consistent with the view that γ-phosphate from ATP is liberated at a special location within the grana at which it precipitates with calcium. This precipitation would account for the electron-opaque deposits seen with the electron microscope under conditions favouring the uptake of calcium, strontium or barium (Nobel and Murakami, 1967).

This light-induced, magnesium-requiring ATP-ase, enhanced by sulphydryl groups, appears to be affected by the hydrogen-ion concentration gradient across the grana membrane. Though triggered by light, this ATP-ase has a pronounced lag before the steady state rate is reached, resembling the similar time-lag in the build-up of steady-state, light-induced proton influx. Both seem to decay with about the same half-time. Further, a hydrogen-ion concentration established by incubating chloroplasts in an acid medium and then transferring them to an alkali medium brings about an ATP hydrolysis that apparently involves this same ATP-ase. Since the divalent cation uptake is inhibited by uncouplers which are probably proton carriers, Nobel suggests that they may inhibit the ATP-ase by preventing the establishment of a hydrogen-ion concentration gradient across the grana membranes.

The system investigated by Nobel is clearly different from that described by Packer, Deamer and Crofts. One of its values will lie in showing the place in chloroplast membranes where the ATP-ase occurs but the relation of this ATP-ase to the over-all mechanism of photosynthesis and to photophosphorylation remains to be shown by future work.

Mitochondrial ion transport

By the end of 1950 I had decided that the accumulation of ions in non-green plant cells must somehow be linked to the mitochondrial liberation of energy (Robertson, 1951). This conclusion was based on the knowledge that the stages in respiration probably concerned (those associated with the cytochromes) were in or on the mitochondria. Further, I liked the idea that the lipid organelles might act as temporary accumulating centres for ions, particularly as they are moved in protoplasmic streaming and would perhaps discharge their ions through other lipid membranes, similar to that at the junction between vacuole and cytoplasm. These views started us on a long investigation of mitochondria themselves. Our colleagues interested in the gastric mucosa, particularly Davies and his collaborators, started similar investigations. Our early work with the electron microscope and the techniques then available had suggested that the mitochondrial membrane was about 200 Å thick (an overestimate) and that it was semi-permeable (Farrant, Robertson and Wilkins, 1953). Subsequently, with the refined section cutting techniques available, we were able to show as others had done that our plant mitochondria had the structure now generally recognised as typical, with a relatively smooth outer membrane and a much convoluted inner membrane (Farrant, Potter, Robertson and Wilkins, 1956). While this work was going on we were also engaged in extensive investigation of electrolyte balance in mitochondria.

The kinds of experiments one does are the products of the hypotheses in mind when they are planned. Our hypothesis was that a considerable quantity of indiffusible negative ions in the particle would be balanced by mobile cations, as had been shown for other types of mitochondria (Bartley and Davies, 1952, 1954; MacFarlane and Spencer, 1953; Stanbury and Mudge, 1953). Further, Bartley and Davies had shown that phosphate and organic acid anions were concentrated in mitochondria. But in addition, we suggested that mobile anions like chloride which are not directly concerned in metabolism might be actively transported into mitochondria by a mechanism associated with the cytochrome system. If, as has been discussed in preceding chapters, mitochondria can secrete protons on the outside and form hydroxyl ions on the inside, the movement of chloride might be either (1) inwards if it exchanged for the hydroxyl ions built up on the inside, or (2) outwards if it were pumped by an anion pump compulsorily coupled to the electron transport inwards, as originally suggested in 1948. Our conclusions from the work of the 'fifties must be reconciled with or reinterpreted by the considerable information now available.

Our early work is summarized as follows:

(1) Cations K^+ and Na^+ were maintained in high concentrations in the mitochondria though they were mostly easily exchangeable. Thus most of the ions could be accounted for by a Donnan system but the remainder, liberated from the mitochondria only by drastic treatment with trichloracetic acid, could not (Honda and Robertson, 1956).

(2) External potassium easily replaced internal sodium but external sodium could not replace internal potassium.

(3) Of all cations, calcium occurred in the mitochondria in much the greatest concentration unless EDTA had been added to the extraction medium; calcium was not replaced by either sodium or potassium. Typical results were as in table 2.

(4) Chloride ion concentrations in the mitochondria were always in excess of those expected on a Donnan equilibrium assuming that the excess sodium and potassium ions were balanced by the indiffusible ions of the particle.

(5) The excess chloride concentration above that of the Donnan expectation was significantly correlated with the oxygen uptake when endogenous substrate was being respired. No correlation between the increased oxygen uptake when substrate was added and the chloride content, was obtained (Robertson, Wilkins, Hope and Nestel, 1955).

(6) By measuring the chloride concentration in the mito-

chondria at different times after changing the external concentration of chloride (the osmotic pressure being maintained with sucrose solution), a half-time of adjustment was obtained and used to calculate the permeability of the mitochondrial membranes as 5×10^{-8} cm/s.

(7) The concentrations of chloride maintained above those expected by the Donnan calculation were approximately those which could be due to the electron-transport system putting ions in, against the back diffusion through the mitochondrial membrane; this conclusion was based on two assumptions, (1) that one Cl^- was moved in for one e^-, passing to oxygen in the

TABLE 2. *Internal concentration of ions in beetroot mitochondria compared with concentrations in the external solution* (mM); *from Honda and Robertson* (1956)

Ion	No EDTA		EDTA	
	External concentration	Mitochondrial concentration	External concentration	Mitochondrial concentration
Ca^{2+}	0·8	404·0	0·4	29·5
K^+	8·3	41·0	10·9	24·6
Na^+	10·3	16·3	7·0	55·0
Cl^-	9·8	7·6	9·5	5·7

respiration, and (2) that the back diffusion would be calculated from the permeability and the average concentration of Cl^- in the mitochondria.

Not much interest seemed to be taken in the ionic balance of plant mitochondria and not much more in that of animal mitochondria until the early sixties. Then from 1961 onwards, following the discoveries of the entry of divalent cations in the presence of phosphate, there were many investigations of this phenomenon (Vasington and Murphy, 1961, 1962; De Luca and Engström, 1961; De Luca, Engström and Rasmussen, 1962; Brierley *et al.* 1962; Brierley, Murer, Bachmann and Green, 1963, and many more since); the principal results can be summarized:

(1) Divalent cations in the presence of phosphate are absorbed by mitochondria.

(2) This absorption occurs either in the presence of substrate supporting oxygen uptake or in the presence of ATP.

(3) Hydrogen ions are secreted simultaneously and the ratio

of monovalent ions absorbed to H^+ ions secreted is 1, whereas that of divalent ions absorbed to H^+ ions secreted may range from 0·5 to 5 and seems to depend on the anion being taken in simultaneously (see Pullman and Schatz, 1967).

(4) Ion uptake occurs only when ATP formation is prevented either by omitting ADP from the medium or by adding oligomycin; thus ion uptake with proton extrusion is an alternative to ATP formation.

(5) Uncouplers and inhibitors of electron transport inhibit ion uptake.

(6) Cation to oxygen uptake ratios approximate to 6.

(7) Though divalent cations are taken up with phosphate alone, intact mitochondria are relatively impermeable to monovalent cations. Permeability to monovalent cations is increased by gramicidin or valinomycin.

The massive uptake of divalent cations and accompanying phosphate absorption is due to the formation of insoluble metal phosphates within the mitochondrion. The evidence for precipitation is: (1) the divalent cation/phosphate ratio approximates 1·5, which is that expected if insoluble phosphates were formed, (2) electron micrographs show deposits of electron-dense material which is presumably the metal phosphate, and (3) neither the cation nor the phosphate leaks from the mitochondrion.

These observations can be explained by the principles outlined at the beginning of the chapter. The extrusion of protons and the inferred alkalinity inside the mitochondria would be consistent with the cations entering in exchange for the hydrogen ions and, if nothing else happened, balancing the hydroxyl ions. However, as phosphate is known to penetrate the mitochondrial membrane, presumably by some kind of exchange diffusion with the internal hydroxyl ions, the phosphate and divalent cations in the alkaline internal environment precipitate as the insoluble metal phosphate. Inhibitors of electron transport prevent the production of protons and hydroxyls, and uncouplers result in the protons being transferred back across the membrane so that no exchange gradient is established; thus both inhibitors and uncouplers prevent the accumulation of the phosphate. This spate of work on accumulation of divalent phosphates in mitochondria was largely because a sufficient quantity was accumulated to analyse easily and to observe with the electron microscope. It taught us various things which may be summarized: (1) that ion movement associated with oxidation was taking place; (2) that this ion movement was an alternative to phosphorylation, occurring only in the absence of ADP or in the presence of ADP with oligomycin; (3) that not all cations were equally effective in entering the mitochondria.

Different workers have shown that calcium, magnesium, manganese and strontium are all taken up accompanied by phosphate, but that different types of mitochondria may show differences in specificity of this uptake; for example, Carafoli, Rossi and Lehninger (1964) showed that Mg^{2+} uptake by rat liver mitochondria was different from Ca^{2+} uptake, which seemed to be particularly favoured in this type of uptake and stoichiometrically related to the electron flux. Subsequently Azzi, Rossi and Azzone (1966) showed that Mg^{2+} affects the permeability of the mitochondrial membrane to monovalent cations. Millard, Wiskich and Robertson (1965) showed that red beetroot mitochondria absorbed Mg^{2+} by an energy-linked process and accumulated phosphate but the process was inhibited by the presence of monovalent cations, though they were not simultaneously accumulated in quantity; monovalent phosphates are soluble and would leak out through the mitochondrial membrane.

The most striking specificity in entry of cations is the difference between calcium, which seems to enter easily, and the smaller monovalent cations to which the membrane appears normally to be impermeable. The first evidence for this is from the experiments of Amoore and Bartley (1958), and Amoore (1960) showed that $^{42}K^+$ exchanged only very slowly with the endogenous K^+ of the rat liver mitochondria—a result rather reminiscent of that which Honda and I obtained in trying to replace endogenous K^+ with Na^+. The later evidence depended on the discovery by Pressman (1963) that the polypeptides, valinomycin and gramicidin, would release respiratory control and result in a simultaneous release of protons and uptake of potassium by mitochondria which had previously appeared to be impermeable to potassium. Chappell and Crofts (1965) confirmed Pressman's results and showed that the normal mitochondrial membrane will become permeable to Na^+ and K^+ in the presence of gramicidin and to K^+ in the presence of valinomycin. Further they have shown (Chappell and Crofts, 1966) that the same antibiotics increase the K^+ and Na^+ permeability of artificial lipid membranes. Chappell and Crofts (1965) showed that gramicidin caused the liberation of H^+ ions by liver mitochondria in the presence of K^+, Na^+, Li^+, Rb^+ or Ca^{2+}. If phosphate was also present, swelling and increased oxygen uptake also occurred. Mitchell and Moyle (1967) showed that rat liver mitochondria, in the presence of valinomycin or gramicidin in a potassium chloride medium, gave a stoichiometric release of protons when the monovalent cation was being taken up.

All these findings pointed to the impermeability of the mitochondrial membrane and, even when the permeability to K is increased by valinomycin, the membrane remains impermeable to protons. Using ^{42}K, Chappell and Crofts (1966) have shown that lecithin artificial membranes treated with valinomycin will allow a very rapid exchange of internal and external potassium. They also showed that gramicidin A and the uncouplers DNP and CCP make lecithin membranes permeable to protons. With mitochondria, Chappell and Crofts (1966) carried out an experiment which demonstrates that gramicidin probably acts by making their membranes permeable to both hydrogen and potassium ions but valinomycin makes them permeable to potassium and leaves them much less permeable to hydrogen ions. If mitochondria were incubated in a choline chloride medium containing succinate under anaerobic conditions, a small efflux of hydrogen ions was observed when oxygen was added. Addition of gramicidin or of valinomycin after this resulted in a marked alkaline shift in the external medium, accompanied by loss of K^+ from the mitochondria. If oxygen was added then, a rapid ejection of hydrogen ions from the mitochondria was followed by re-entry but the re-entry was more rapid in the presence of gramicidin than in the presence of valinomycin.

All these results are consistent with a mitochondrial membrane which is relatively impermeable to hydrogen ions and monovalent cations but which is penetrated more readily by divalent cations. Since the divalent cations are less likely to penetrate simple physical pores than are the monovalent cations, some kind of combination in the membrane with liberation of the cations on the inside must be postulated. The

49

idea is that the membrane is capable not only of generating protons on the outside and hydroxyls on the inside but also of 'choosing' the ions that are moved as the result of the potential difference set up.

The capacity of mitochondrial membranes to 'choose' rather specifically what ions cross them is strengthened by the remarkable results which have been obtained with anions. Much of this work has been done by Chappell and collaborators and is summarized in their most recent publication (1967). The striking results are perhaps best expressed by saying that the mitochondrial membranes which are relatively impermeable to the small chloride ions are quite readily permeable to such large ions as di- and tri-carboxylic acids, glutamate, aspartate, phosphate, ADP and ATP. Table 3 taken from Chappell, Henderson, McGivan and Robinson (1967) summarizes the results which have been obtained for different anions and by various investigators.

Chappell and Crofts (1966) had used a simple technique for investigating the rate of penetration by anions. It had been found that mitochondria suspended in isotonic solutions of certain ammonium salts will swell. It is probable that ammonia enters in the form of NH_3 and, on the inside of the mitochondrion, associates with a proton to give NH_4^+ which cannot diffuse out as readily as the OH^- formed simultaneously. This hydroxyl ion therefore exchanges with any anion from the external solution to which the membrane is permeable. The entry of the anion to balance the ammonium results in an increase in internal osmotic pressure and the mitochondria swell. In solutions containing the chloride, bromide or sulphate of ammonium, the mitochondria do not swell but they do so in ammonium arsenate, phosphate or acetate. They do not swell in K^+ or Na^+ salts of arsenate, phosphate or acetate.

This technique has been used by Chappell and Haarhoff (1967) to investigate the entry of other anions. Formate, propionate and butyrate entered rapidly. With dicarboxylic acids, little or no swelling occurred unless phosphate or arsenate was added simultaneously in low concentration (e.g. 2 mM-NH_4 phosphate to 100 mM-NH_4 succinate). D-malate, L-malate, methylene succinate, malonate and mesotartrate also required the phosphate—referred to as an activator (table 3). This phosphate-dependent swelling did not occur with fumarate,

TABLE 3. *The transporting systems of mitochondria*

Substance transported	Activators	Inhibitors	Occurrence in mitochondria
1. Phosphate, arsenate	None	—	Liver, kidney, brain, heart, etc., probably all mitochondria
2. Some dicarboxylic acids, e.g. malate	Phosphate	Butylmalonate	All mitochondria except blowfly flight-muscle
3. 2-Oxoglutarate	Phosphate + malate, malonate, etc.	Butylmalonate	Liver, brain, heart-muscle, kidney, *Candida utilis*, *not* blowfly flight-muscle
4. Citrate, isocitrate, cis-aconitate	Phosphate + malate, 2-hydroxymalonate, 2-hydroxy-2-methylmalonate	Butylmalonate	Liver, kidney, low activity in heart and brain. Absent in blowfly flight-muscle
5. Glutamate	Phosphate (slightly)	4-Hydroxyglutamate, L-amino-adipate, *three*-DL-hydroxy-aspartate	Probably present in all mammalian mitochondria
6. Aspartate	Glytamate, 4-hydroxyglutamate, L-amino-adipate, *three*-hydroxyaspartate	—	Shown in liver, kidney, heart, probably present in all mammalian mitochondria
7. ATP, ADP	Mg^{2+} ?	Atractyloside	Probably all mitochondria

maleate, citraconate, mesaconate, D-tartrate or L-tartrate as the anions. This suggested that these anions could not penetrate and that, not only is an activator required, but also that the anion should have a particular molecular configuration 'acceptable' to the membrane.

When citrate and cis-aconitate were the anions being investigated, not only phosphate but also L-malate (again in relatively low concentration) was necessary; none of the other dicarboxylic acids could replace the L-malate but its effect is competitively inhibited by butylmalonate.

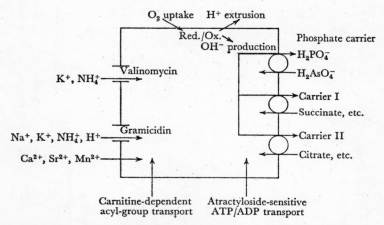

Fig. 12. A scheme depicting ion movements in liver mitochondria from Chappell and Haarhoff (1967).

These observations make important contributions to our speculations on the nature of the membrane and the processes taking place in it. Chappell and Haarhoff considered it unlikely that the unionized forms of the monocarboxylic acids enter rather than a process involving OH^-/anion exchange, because formate (little unionized at pH 7) enters as rapidly as acetate (much unionized at pH 7). With dicarboxylic acids, the proportion unionized is very small and they suggest that a singly or doubly charged anion probably penetrates. There is no correlation between amount of singly charged anion and the apparent rate of penetration. Chappell and Haarhoff also discuss the evidence, from work on utilization of some of these anions, that their rates of penetration might depend on the specificity of the membrane. This is not the place to review all the evidence or speculate on the mechanism which might be concerned in the specific entry. It is sufficient for our purposes to note that the evidence is consistent with three of our basic postulates: (1) that the membrane is relatively impermeable to even small ions such as hydrogen and hydroxyl ions, (2) that oxidation/reduction reactions result in the separation of charge, (3) that the ion exchange mechanisms then possible are regulated by specific substances in the membrane which will allow the exchange of *some*, but not *all* anions for the hydroxyl inside.

It is useful to summarize the present knowledge of these ion uptake properties of mitochondria and fig. 12 is modified from that of Chappell and Haarhoff (1967); the main modification is that the term 'hydrogen pump' is avoided because it seems more appropriate to talk of hydrogen extrusion and accompanying hydroxyl production.

Swelling and shrinking

Considerable work has been done on the swelling and shrinking properties of both mitochondria and chloroplasts; no full discussion of all the variants of behaviour will be attempted here. It is important to note that not all the work which has been done leads to reliable interpretations. Many investigations, for instance, have depended on light-scattering techniques, assuming that the greater the light scattered, the smaller the particle size. While this is generally true over a range of particle size, the results should not be accepted as indicating merely a change in volume unless the light-scattering changes have been correlated with volume changes measured by one or other of the alternative methods. Each of the methods has some disadvantage, but where appropriate, Coulter counter, haematocrit absorbancy decrease and electron-microscope structure can each be used. Light scattering has the advantage that it is quick to do so, does not destroy the particles being investigated and can trace changes in time. The work of Packer and his colleagues has given due consideration to critical comparison of the methods and should be consulted for details. The important paper by Hackenbrock (1966) shows how internal mitochondrial structure may be changed (as observed by electron microscopy) without corresponding change in over-all volume of the particle.

53

The mitochondrial behaviour can be illustrated by the work of Azzone and Azzi (1965), who observed swelling and shrinkage cycles reversible by the addition of EDTA. The swelling phase was dependent on energy supply—either from the respiratory chain or from ATP. Inhibitors and uncouplers initiated a shrinkage phase. In 1966, Azzi, Rossi and Azzone showed that this swelling in the presence of EDTA is coupled to a respiratory-dependent cation uptake by the mitochondria; the shrinkage phase is accompanied by a respiration-independent cation diffusion outwards. Swelling was greater in lithium chloride than in sodium chloride, in which swelling was greater than in potassium chloride. Inorganic phosphate induced more swelling than chloride but organic anions were also effective, especially if accompanied by lithium. The swelling phase was accompanied by a large hydrogen-ion ejection and an uptake of univalent cation and increased respiration. 2,4-DNP reverses the swelling; that is, causes contraction and H^+ ions pass back into the particle. These observations can be interpreted as being basically due to the separation of OH^- and H^+, with cations moved in at first, anions exchanged for OH, osmotic pressure increased and water entering, resulting in swelling. If the respiration is inhibited or uncoupled, the gradient is not maintained and cations and anions already there gradually leak.

A specially interesting oscillatory state of mitochondrial swelling and shrinking was investigated by Packer, Utsumi and Mustafa (1966). Taking EDTA-treated rat liver mitochondria, they were able to obtain a damped oscillation in swelling and shrinking with a periodicity of about 2 min (see fig. 13) but this period depends on pH and temperature. The system requires a high pH; EDTA; absence of Mg^{2+}, Ca^{2+} and Sr^{2+}; an energy source either from electron transport or from ATP; a permeant anion which may be phosphate or acetate or propionate; appropriate conditions of osmolarity; oxygen uptake. The process can be initiated by adding one of the essential constituents last; substrate, ATP, the permeant anion or the oxygen, so added will start the reaction. During the swelling period there is rapid production of protons which decreases during shrinkage. Shrinkage is associated with more highly oxidized state of pyridine nucleotides, stimulation of ATP hydrolysis, inhibition of proton extrusion and increased cation efflux.

What starts this interesting phenomenon and how is it

54

related to the primary act of hydrogen and electron separation? According to the hypothesis developed early in this chapter, the availability of an anion would be essential for the initial charge separation. In the system with everything present except anion, no oxygen uptake would occur. On addition of the permeant anion, the electron transport starts, H^+ is extruded with the

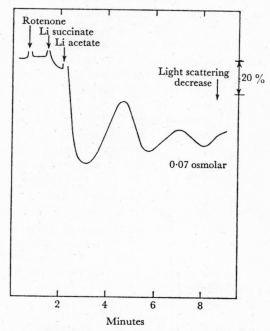

Fig. 13. Oscillatory swelling of mitochondria measured as decrease of 90° light scattering at 650 mμ (from Packer, Utsumi and Mustafa, 1967).

permeant anion transported from inside to outside, the electron passes inwards to form OH^- with oxygen and water and the cation moves in balancing the OH^-. The swelling which occurs concurrently is probably due to a combination of pH change on the inside affecting structure, entry of weak electrolytes from the acid exterior to the alkaline interior where they dissociate and, with the entering cations, increase the internal osmotic pressure. After the swelling of the mitochondrion has reached a certain point, the stretched membrane probably becomes permeable, water and ions leak to the exterior and the

E

55

mitochondrion shrinks. After a time the cycle can start again with hydrogen extrusion and swelling. It should be noted that a membrane which is permeable to the anion or the undissociated molecule of a weak electrolyte, would allow this cycle to proceed without much build-up of anion concentration on either side. If the membrane is impermeable to anion but permeable to the undissociated weak electrolyte, some build-up of the anion concentration on the alkaline side of the membrane could occur, but this would depend on the relative rates of diffusion of the undissociated molecule and of transport of anions; further, the alkalinity would be reduced by the hydrogen ions from the weak electrolyte and the net result would be electron transport and oxygen uptake with no marked build-up of alkalinity, acidity or ion concentrations.

With a strong electrolyte like KCl, supplied to mitochondria with endogenous substrate, the rates of oxygen uptake could be limited by low concentrations of anion inside the mitochondria, as the electron transport would lack sufficient anions for the coupled transport. This hypothesis might explain the correlation between internal Cl^- concentration and oxygen uptake in our beet mitochondria (Robertson *et al.* 1955) with no substrate supplied and with low external KCl. By contrast, no correlation between internal Cl^- and oxygen uptake was obtained when substrate was supplied; as all such substrates are weak electrolytes, no shortage of anions to activate the electron-transport system would be expected.

The nature of the membranes

Despite much work on membranes, we have yet to learn how the critical enzyme systems are orientated in mitochondrial and chloroplast membranes to bring about the functions discussed in this chapter. Work on mitochondrial membranes has been reviewed by Parsons (1966) and recent developments are outlined by Pullman and Schatz (1967). I shall not attempt detailed speculation on the relation of hypothesis of structure to the hypothesis of function developed in this book. We know that the inner mitochondrial membrane, which carries out electron transport and oxidative phosphorylation, has a complex structure with knobs of about 90Å diameter on the inside. These knobs have been shown to contain the oligomycin-sensitive ATP-ase, and the membrane proper to contain the associated electron-transport system. Together these make the structure which can bring about charge separation, phosphorylation and ion movement. The two sides of the mitochondrial membrane are different both in function and form. This difference has been demonstrated clearly by the elegant experiments of Mitchell, Moyle and Lee (see Mitchell, 1966); isolated mitochondrial membrane fragments round off into vesicles, so these workers were able to demonstrate proton extrusion by the vesicle when the outside of the membrane was to the outside of the vesicle and proton uptake when the outside of the membrane was to the inside of the vesicle.

Conclusion

In this chapter, I have emphasized that membranes exemplified by the mitochondrial inner membrane and by the lamellae of chloroplasts under the conditions outlined are pre-eminently suited to: (1) separating proton and electron; (2) simultaneously moving an anion as a process coupled to, but opposite in direction to, the electron movement; (3) keeping H^+ and OH^- separated long enough to allow traffic in other ions; (4) discriminating between different ions in exchange diffusion.

In the next chapter I shall discuss how these properties of membranes of organelles and probably of other cell membranes may be of great importance to the cells themselves.

CHAPTER 5

Charge Separation, Active Transport and Control in Cells

'Theories are only hypotheses, verified by more or less numerous facts. Those verified by the most facts are best; but even then they are never final, never to be absolutely believed.' (CLAUDE BERNARD, translated by H. C. GREENE)

THIS book has been written to put a point of view; that is, to suggest reasonable explanations based on the idea of charge separation which might be a unifying hypothesis (Robertson, 1960, 1967). As pointed out earlier, most of the examples discussed are from my own field of active transport in plant cells, but I believe that the underlying principles will be similar for some of the kinds of active transport seen in animal cells. Before considering the implications and limitations for the cell as a whole, it is useful to review briefly some of the earlier hypotheses about the ways in which plant cells take up ions in active transport; somewhat similar hypotheses have been suggested from time to time for active transport by animal cells. How can the plant cell take ions from the external solution and, passing them through the cytoplasm with its characteristic outer membrane (the plasmalemma) and inner membrane (the tonoplast), accumulate them by active transport in the vacuole?

Older hypotheses

(a) Hypothesis of exchange for H^+ and HCO_3^-

Brooks (1929) suggested that the continuous production of H^+ and HCO_3^-, as the result of CO_2 liberation in respiration, might set up a system to allow an exchange of a cation for the H^+ and of an anion for the HCO_3^-. This mechanism, somewhat analogous to the charge separation hypothesis, which would not work in the form originally suggested by Brooks (1929) but would work in the modified form suggested by

Briggs (1930), will not account for the ratio of 4 g equiv. of salt accumulated to 1 g mol. O_2 absorbed. The maximum rate of accumulation on this hypothesis would be when 1 g equiv. of cation was exchanged for 1 g equiv. of H^+ and 1 g equiv. of anion for 1 g equiv. of HCO_3^-. Since 6 CO_2 molecules are produced for 6 oxygen molecules taken up, the salt/oxygen ratio would not exceed one.

(b) *Hypothesis of direct relation to protein synthesis*

In the early 1930s, Steward had shown how important oxygen uptake was in relation to salt accumulation by a number of different tissues. He also found that a number of tissues would absorb ions only if they had simultaneously a capacity for growth. Steward and Preston (1941) showed that salt uptake was accompanied by protein synthesis in potato tissue and went on only if protein synthesis was taking place. The hypothesis was that growth and salt accumulation both depended on protein synthesis but no mechanism for connecting the two was suggested. In 1947, Steward and Street suggested that energy-rich phosphorylated nitrogen compounds might function as ion carriers. In 1948 when I had done much of my quantitative work with carrot tissue, I remarked in conversation with Steward that I thought carrot tissue had great advantages for investigating the problem of accumulation (which we know now as active transport) because it was a non-growing tissue with a good salt accumulation. I remember his quick retort: 'Yes, but for me that's like Hamlet without the Prince of Denmark.' I think that conversation summed up our difference of approach, for whereas he was interested in finding out about uptake of all the ions entering during development of cells, I was interested in isolating the active transport process if possible, to investigate by itself. Subsequently Steward and his collaborators used carrot tissue for their fine work on growth factors, particularly those in coconut milk. Comparing tissue cultures of dividing and growing cells (with coconut milk factors) with cultures of vacuolated non-growing cells (like my carrot slices with no growth factors), they found active transport carried out more by the non-growing cells than by the growing (Steward, 1954). This strengthened my view that neither growth nor protein synthesis was directly necessary for active transport. Needless to say, the growing cells took up ions as required by their metabolism and increased in cation content by exchanging for the protons from the metabolic anions which increase in growing cells, but active transport was most developed after vacuolation.

In the late 1950s, the protein synthesis hypothesis had a revival because of Sutcliffe's observations (1960) that chloramphenicol, which depresses protein metabolism in bacteria, inhibited active transport of ions in beetroot cells. Chloramphenicol had its effects on active transport without much effect on oxygen uptake; this led Sutcliffe (1962) to postulate that the active transport mechanism was directly dependent on protein synthesis. Chloramphenicol inhibited influx by about 50% at once and inhibited net uptake completely after about 8 h, primarily because the tissue in the presence of the inhibitor is saturated with electrolyte at a lower level than that in the absence of inhibitor. E. A. C. MacRobbie (personal communication) has confirmed the inhibition of tracer influx. However, the difference between respiration rates in salt and in water is down by about 50%. Since this respiration may be critical, it is not valid to conclude that the uptake is linked directly to protein synthesis or that respiration is not involved. Recently the evidence for the chloramphenicol effects has been examined critically by McDonald, Bacon, Vaughan and Ellis (1966), who conclude that their results do not support the view that ion uptake is directly linked to protein synthesis. In putting forward a hypothesis to relate protein synthesis to ion uptake, Sutcliffe (1962) developed the idea suggested by Bennett (1956) that the formation of vesicles by living membranes may be part of the mechanism of ion movement and active transport. We shall return to this idea later.

(c) Protein folding and unfolding

Goldacre (1952) advanced an ingenious and plausible hypothesis based on observations of the accumulation of dyes in *Amoeba* and in root hairs. He suggested that a cyclic folding and unfolding of proteins in which polar linkages (ionic bonds) were alternatively made and broken would create a succession of positive and negative charges on the protein which would cause local concentrations of anions and cations respectively. When the proteins are in the extended state, anions would be adsorbed on the positive groups and cations on the negative.

When the protein passes to the contracted state, the positive and negative groups compensate each other intra-molecularly and the ions are desorbed. Goldacre demonstrated that dyes pass into the vacuoles of root hairs at the same time as protoplasmic streaming and as rhythmical migration of vacuole from one end of the root hair to the other and back again. He connected these processes with the hypothesis that all are dependent on protein folding and unfolding but, as far as the evidence goes, other interpretations are possible and the proof that folding of proteins has any connexion with the ion accumulation mechanism is lacking. Goldacre's idea may be important in some contexts, though it seems unlikely to be the general phenomenon by which energy generated by respiration or photosynthesis is transferred to the process of ion accumulation.

(d) Carrier hypotheses based on kinetic experiments

Many experiments on many kinds of tissue have now shown that over certain ranges of concentration the relation between the reciprocals of external concentration and of the rate of ion absorption is approximately linear. This relation to concentration is the same as that observed in enzyme kinetics where it indicates that the substrate can combine with the enzyme (which is present in small amounts) until the enzyme is saturated with substrate and the maximum rate for that concentration of enzyme is approached. The relation is

$$v = \frac{Va_0}{(a_0 + K_m)},$$

where v = velocity of the reaction (or of active transport), V = the maximum velocity approached as a_0, the concentration of substrate (or of salt in active transport) increases, K_m = a constant which often approximates to the dissociation constant of the enzyme–substrate complex (or of a hypothetical carrier of an ion in active transport).

This type of analysis, originally developed by Epstein and Hagen (1952), has been applied with great precision, particularly by Epstein and his collaborators. Some of the difficulties of interpretation of these results have been discussed by Briggs, Hope and Robertson (1961), but in general studies of this type have added important information about the behaviour of cells to different ions. The results are consistent with the view that

61

certain combining sites in the cell membrane or in the cytoplasm become 'saturated' with the ion under investigation at
certain concentration. Further, the important observations of
Epstein, Rains and Elzan (1963) have shown that the uptake
of ions by barley roots does not follow a smooth curve over the
whole range of concentrations and that different mechanisms
operate at low and at high concentrations. The results of this
work have recently been reviewed by Epstein (1966) and a
typical curve is shown in fig. 14.

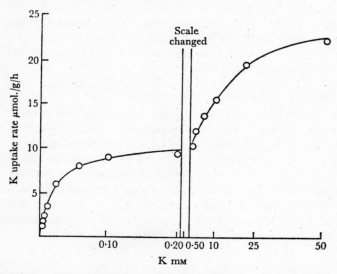

Fig. 14. Rate of absorption of potassium labelled with ^{42}K as a function of the
concentration of potassium chloride. The concentration scale is changed between
0·20 and 0·50 mM. (From Epstein, 1966.)

Another valuable product of this work has been its contribution to defining better the specificity of the uptake in which
different ion species in the external solution compete, or do not
compete, with each other for entry into the cells. From these
observations arose a simple carrier hypothesis; that is, that the
hypothetical substance, which became saturated with the ion
in question, was a carrier which combined with the ion at the
cell surface and liberated it on the inside. We have already
seen that there is strong evidence for the combination of certain
ions quite specifically in the mitochondrial membrane, which
from some points of view gives greater plausibility to the carrier

hypothesis. Further, we should be looking for mechanisms of active transport (not for a single mechanism). It is relevant to note, for example, that Pitman and Saddler (1967) with barley roots have evidence that at low concentrations active transport may be controlled by a sodium–potassium pump. At higher concentrations another mechanism operates (possibly an anion pump). We shall return to these problems. Though it is important to note that no carriers such as this hypothesis requires have been isolated, this could well be an almost hopeless task for the carrier substance may have no chemical existence separated from the structure of the membrane of which it is a part.

(e) *The lipid-soluble carrier*

Various people have suggested that any such carriers must be compounds which take ions across lipid membranes, and the last hypothesis to be mentioned is an example which makes use of knowledge of the characteristics of plant cells. Bennet-Clark (1956) was impressed with the similarity of indole-acetic acid to certain substances which inhibit choline-esterase activity in animal cells, and with the observation that 10^{-3}M indole acetic acid causes a 50–60% inhibition of salt uptake by beet tissue. He sought a hypothesis to connect these two facts in the properties of lecithin which is both a fat-soluble substance and a zwitterion. Such a substance might combine with an anion and cation at a membrane surface which the free ions themselves would not be able to penetrate. If, on the inside of the membrane, lecithin were hydrolysed by lecithinase-D to choline and phosphatidic acid, the salt would be liberated on that side of the membrane. It was suggested then that choline might be converted to acetylcholine by the acetylase system (the energy being derived from ATP), and choline esterase then might transfer the choline from the acetyl to the phosphatidyl radical, thus resynthesizing the lecithin at the outside of the membrane (fig. 15). Similar hypotheses have been suggested for active transport in some animal cells but, in common with most carrier theories, evidence to support these hypotheses is lacking; the value in the present context is to make the point that various hypothetical chemical mechanisms can be suggested on the reasonable assumption of (1) combination with the ion or ions, (2) different behaviour on the two sides of the membrane,

(3) lipid characteristics of the membrane, impermeable to the hydrated ions once liberated from the carrier.

This survey—by no means exhaustive—of some of the older theories of active transport brings us to the apparent paradox of ion transport by cells. On the one hand, there are apparently only two basic sources of energy, respiration and photosynthesis, which are localized in two parts of the cell (mitochondria for aerobic respiration and chloroplasts for photosynthesis),

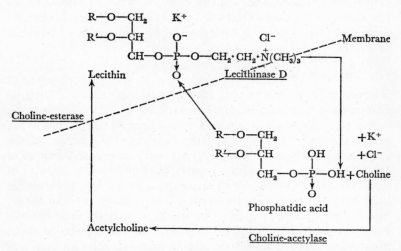

Fig. 15. The mechanism of ion uptake suggested by Bennet-Clark (1956) with a postulated phosphatide cycle as a carrier mechanism.

but on the other hand many ions are transported actively, some at different membranes of the cell—organelle, plasmalemma and tonoplast. Further, the different mechanisms can distinguish different ions, even in different parts of the same cell. In the rest of this chapter we shall be concerned with the general principles based on our unifying hypothesis of charge separation but allowing for some of these differences.

Kinds of pumps

At this stage it is useful to summarize the kinds of active transport which are now known to occur in living cells, by referring to the type of pump responsible. As pointed out in the beginning of chapter 2, only one ion of a salt or ion pair needs to be moved against the electrochemical potential gradient; the second ion of the pair will follow the one moved. If the anion is moved by expenditure of energy on the part of the cell, and hydrogen ions or other cations follow, an anion pump is said to be operating. If a cation is moved and an anion or anions follow, the pump is a cation pump. One special pump of common occurrence results in the extrusion of sodium and the simultaneous absorption of potassium—a sodium–potassium pump, usually referred to as the sodium pump. Further, there is evidence that there may be other pumps which are specific for particular ions; for example, the calcium uptake by mito-chondria.

The problems in explaining energy-requiring pumps are:

(1) The site of the pump—cell surface membrane, endo-plasmic membranes, organelle membranes, vacuolar mem-branes (in plant cells) are all possibilities.

(2) The site of the energy source; as far as we know at present the ultimate energy sources are the glycolytic mechan-ism (e.g. in the red cell), the mitochondria and the chloroplasts.

(3) The transfer mechanism for energy from its source to the site of the pump; as far as we know, there are three ways in which energy can be transferred: (a) transfer of ATP, (b) transfer of reducing power in the form of hydrogen or electron carriers, e.g. NADH, NADPH; (c) transfer of H^+ or of OH^- after charge separation.

(4) The mechanisms of the pumps.

When we consider the various combinations of active trans-port systems which can be expected in living cells, it is clear that we must be looking for mechanisms of active transport and no one hypothesis of 'the mechanism' will fit all the different possibilities. Some suggested explanations for the different pumps will be considered.

Anion pumps

(a) Nature's experiment in the gastric mucosa

It is helpful to consider the special arrangements for the secre-tion of acid and alkali by the gastric mucosa. Here, following the hypothesis outlined earlier, we can recognize two important characteristics:

(1) That the separation of charge is so complete that the

H$^+$ and OH$^-$, or HCO$_3^-$, cannot diffuse together again but leave the cell on opposite sides. Further, cells are equipped with membranes which can guide acid (0·13M-HCl) to the lumen of the stomach with pH as low as about 1 and other membranes which provide for the alkali to pass back to the blood stream on the other side of the cell.

(2) That the chloride is primarily transported; that is, it is a true anion pump. The evidence is (a) that the chloride concentration (about $1\frac{1}{2}$ times that of the blood) exceeds the hydrogen-ion concentration, the balance being made up of K$^+$ so it is unlikely that the H$^+$ transport is the primary act, and (b) the more critical evidence from measurements of electrical potential and chloride flux which show that the chloride is not moving along the electrochemical potential gradient (Hogben, 1955).

In my view the observations on the mucosa have several important consequences which might be applicable to other cells as well:

(1) The H$^+$ secretion and OH$^-$ diffusion into the blood are probably due to the separation of proton and electron at the electron-transport system.

(2) This charge separation is accompanied by an anion (chloride) pump due to the compulsory coupling of the anion movement to the mechanism for separating proton and electron.

(3) All this could be happening at the mitochondrial membrane.

(4) The ions, once separated—possibly to different sides of the mitochondrion as far as our present evidence goes—are kept apart by membranes which are not destroyed by high acidity or alkalinity on one side. The complex structure of the cell of the gastric mucosa as revealed by electron microscopy is consistent with the formation of microtubules or vesicles. Some of these vesicles might fill with HCl and discharge in one direction (towards the lumen of the stomach) and others might fill with sodium bicarbonate and discharge in the other direction (towards the blood). If one type of cell is organized in this way to carry out a special function, the same kind of mechanism separating acid and alkali may be involved in other cells, though it is not so apparent because they are not specialized in acid secretion. The possibility is that such a separation of

66

hydrogen and hydroxyl ions may contribute to the transfer of energy from the mitochondrion to other sites, including those for ion transport in other cells.

(b) Anion pumps in plant cells

Like the giant squid axon, giant algal cells have the advantage of large size for manipulation; another advantage is that they accumulate inorganic ions in the large vacuoles, which can be analysed or can have electrodes inserted in them. Electrodes can also be inserted in the cytoplasm. By study of fluxes of ions, of potential differences between the different parts of the cell and of impedance measurements, the properties of some of the membranes and the location of the special pumps which operate on particular ions are well understood. If the cells are at flux equilibrium—that is, the net rate of input of ions equals the net rate of leakage of ions—and if no active transport mechanism is operating, the concentration of an ion inside can be predicted from the outside concentration and the potential difference between inside and outside. Electrodes can easily be inserted in the vacuoles and the differences in potential measured. The relationship is given by the equation $C_o = C_i \exp(ZFE/RT)$, where C_o and C_i are the concentrations inside and outside respectively, Z is the valency of the ion, F is the faraday, R is the gas constant, T is the absolute temperature and E (where $E = E_o - E_i$) is the potential difference between outside (E_o) and inside (E_i). If the observed concentration inside exceeds that calculated, then the ion is being moved against the electrochemical potential gradient and an inward pump is indicated. If the observed value is less than that calculated, then an outward pump is indicated. Figures for the chloride concentration in the vacuoles of a number of algal cells are compared in table 4. Clearly in most of these cells an anion pump seems to operate.

No attempt is made to summarize all the work on giant algal cells but some of the principles can be well understood from MacRobbie's work on *Nitella translucens* (MacRobbie, 1965).

This cell shows an active influx of potassium and an active efflux of sodium, probably at the plasmalemma, and an active influx of chloride which is also into the cytoplasm, probably at the plasmalemma. The evidence is, as usual, the direction of the flux against the electrochemical potential gradient. Energy can be supplied to the active chloride mechanism either by photosynthesis (in the light) or by respiration (in the dark); the chloride influx in light is about 17 times that in the dark. This light-stimulated chloride uptake seems to be linked to the electron flow and to be independent of ATP, thereby contrasting with the potassium uptake which is ATP-dependent (to be discussed later). Increase in electron flow in the dark, due to

TABLE 4. *Chloride and potential differences in giant algal cells*

Cell	Chloride concn. (m-equiv./l.)		Potential difference (mV) between outside and vacuole	Reference
	Medium	Vacuole		
Nitella translucens	1·3	160	− 120	MacRobbie (1962)
N. flexilis	0·7	135	− 160	Kishimoto, Nagai and Tazawa (1965) Kishimoto and Tazawa (1965a)
Hydrodictyon africanum	1·3	38	− 90	Raven (1967)
Lamprothamnium succinctum	350	370	− 100	Kishimoto and Tazawa (1965b)
Valonia ventricosa	596	643	+ 17	Gutknecht (1966)

an uncoupling agent, stimulated the chloride uptake. In an ingenious experiment, using the two wavelengths of light at which the different reactions of photosynthesis occur (fig. 8) and with inhibitors of various steps in the process, MacRobbie was able to show that the uptake of chloride may be close to the light reaction which does not proceed at wavelengths above 705 mμ (light-reaction II). At this part of photosynthesis, Bové, Bové, Whatley and Arnon (1963) have shown that Cl⁻ or Br⁻ is an essential factor. Uptake of chloride also occurred in an atmosphere not containing CO_2 and was inhibited by the low concentrations of DCMU which stop photosynthesis. The uptake of phosphate as shown by Smith (1966) is apparently dependent on a different mechanism, for its absorption is stimulated by light but the dark influx remains at 65–70% of the light value. All these results would be consistent with the suggestion that the potassium uptake mechanism depends on

ATP, that the phosphate uptake mechanism depends on factors at present unknown, but that the chloride uptake mechanism might depend only on the charge separation associated with the light reactions.

One other aspect of the plant cell should be summarized at this stage. The cytoplasm is not a homogeneous structure and electron microscopy shows that a complex system of membranes forms the endoplasmic reticulum. Though the electrical conductivity of the cytoplasm is quite high in the direction parallel to the surface, the electron microscope shows membranes cutting off small compartments within the cytoplasm. We know nothing of the permeability properties of these membranes of the endoplasmic reticulum. These small compartments often take the form of micro-vesicles which might contain substances in different concentration from outside the vesicle and might play an important role in transport of ions. Furthermore the cytoplasm, with this complex system of compartments, must be remembered as a streaming mass, moving inside the relatively stationary outer surface. The plasmalemma is a recognizable structure with the electron microscope and is a high-resistance membrane relatively impermeable to ions. The properties of the plasmalemma in algal cells were first established by Walker (1957). The tonoplast is in motion at the junction between the streaming cytoplasm and the moving vacuolar sap.

An important gap in our knowledge at present is the connexion between the chloroplast (where, as MacRobbie showed, the connexion with light reactions can be demonstrated) and the mechanism of entry of ions at the plasmalemma or passage of ions at the vacuolar membrane (tonoplast). MacRobbie (1964) has obtained evidence that the size of the flux of chloride at the plasmalemma is correlated with the size of the vacuolar flux. One of the several possible explanations is that these two fluxes are both dependent on a common metabolite in the cell or on a common reaction and we must look for a connecting mechanism in the metabolism of the cell. Finally, E. A. C. MacRobbie (personal communication) has shown recently that the rate of entry of chloride to the vacuole of a *Nitella* cell is not steady but discontinuous and behaves as though the total number of ions arriving at any time is about 50. It is difficult to explain this 'package' of ions accumulated at any one time, unless micro-vesicles of certain average size which fill to a certain level are postulated. These vesicles might then be pictured as discharging the ion contents into the vacuole by reversed pinocytosis.

(c) *In non-green cells*

In the cells of roots and storage parenchyma, the ions transported to the vacuole are dependent on respiration as a source of energy. For our present purposes, the behaviour of ions in the free space, water free space, and Donnan free space of the cell wall is not of immediate interest and we can describe the properties and behaviour of the cell from the plasmalemma inwards.

The evidence for the electrical characteristics of the plasmalemma of higher plant cells was obtained recently by Greenham (1966), who demonstrated an average surface resistance of 3000 Ω/cm^2 with cells of cucumber, oats and maize.

Pitman (1964) working with beet tissue and Cram (1967) working with carrot tissue have measured the fluxes of ions by tracer techniques. Both tissues show three distinct rates of adjustment with time: a fast adjustment component which has all the characteristics of the free space; a slower adjusting component which would be consistent with the low permeability at the plasmalemma and is therefore the adjustment of concentration in the cytoplasm or some part of the cytoplasm; a very slowly adjusting component which would be consistent with a low permeability at the tonoplast and is therefore the adjustment of concentration in the vacuole.

As discussed briefly earlier in this chapter, there is evidence first obtained by Epstein and collaborators for an uptake mechanism which tends to saturate at 0·2 mM. Similar evidence has been obtained by Torii and Laties (1966) in corn roots, by Pitman and Saddler (1967) in barley roots and by Cram (1967) in carrot. This system seems to have an active anion uptake at the plasmalemma. When salts are being accumulated from higher concentrations, apparently an anion pump, whose exact location is unknown, works to accumulate the salt in the vacuole (Bowling, Macklon and Spanswick, 1966). This system, which becomes independent of salt concentration only at high concentration, is certainly the same as the one we have investigated in carrot tissue, where it is related directly to the salt respiration (Robertson and Wilkins, 1948 b). The evidence for the anion pump is the comparison of the potential difference between the vacuole of the cell and the external solution with the concentration ratio inside to outside. A survey recently published by Higinbotham, Etherton and Foster (1967) gave measurements from segments of roots and epicotyl internodes of peas, and from roots and coleoptiles of oat seedlings. With the eight major nutrients supplied, all the anions Cl^-, NO_3^-, $H_2PO_4^{2-}$ and SO_4^{2-} were being moved to the interior of the cell against their electrochemical potential gradients, so an active transport mechanism was operating. Potassium movement might have been due to the electrochemical potential gradient, but Na^+, Mg^{2+} and Ca^{2+} must either have been extruded or excluded. Since the anions cannot be there because they are following cations into the cell, an anion pump must be inferred. Where is this pump and what is its mechanism of operation?

We have had for a long time some evidence about the surface of the carrot cell which has puzzled me. Salt respiration was stimulated rapidly when salt was put on the outside of the tissue and active transport began. If, after a period of salt accumulation, the external solution was replaced with water, very little of the salt that had been accumulated was lost. The salt respiration decreased only very slowly, taking about 48 h to drop to about 30% of its maximum. When a new lot of salt was added to the external solution the salt respiration rose rapidly to its maximum again within 20 min (Robertson and Thorn, 1945). I had always interpreted the low loss from the tissue in water as being due to the low net efflux from the cell surface, though there might have been rapid turnover of salt between the vacuole and the cytoplasm. Gradually the salt was moved out of the vacuole and into the cytoplasm. The low net efflux would be consistent with Greenham's observation that the resistance of the tonoplast is only about 20% that of the plasmalemma. Recently important evidence has been obtained by Cram (1967). He took carrot tissue in which the cells had been loaded with labelled chloride (^{36}Cl), transferred it to an unlabelled chloride solution and measured its efflux; this settled to a steady efflux at about 30 counts/min. He then transferred the tissue to water and the efflux dropped almost to zero, showing that exchange diffusion of chloride from the surface to the external solution governed the efflux; if no anion with which the chloride can exchange is supplied to the external

F

71

solution, the efflux is almost zero. If the tissue is now transferred from water to a solution of unlabelled chloride, there is
a rapid increase in efflux rate of labelled chloride for about
30 min followed by a period of decreased efflux until it settles
down to the same steady rate as before (about 30 counts/min).
This transient increase in rate of efflux would be consistent
with labelled chloride from the pool in the vacuole having
exchanged with the chloride, which could not leave the surface
of the cell because of the lack of exchangeable anion in the
water outside. We can thus picture a rapid turnover of chloride
between cytoplasm and vacuole, but loss of a chloride from the
surface of the cell depends on the presence of an exchangeable
ion in the external solution. Recently W. J. Cram (personal
communication) has found that labelled chloride will leave the
surface of the cell if bromide or iodide is available in the external solution for exchange but only to very limited extent if
nitrate is the anion in the external solution and not at all if
sulphate is the anion. The most probable explanation of these
results is that the plasmalemma, which is very impermeable to
ions in general, contains a compound with which chloride and,
to some extent, iodide and bromide can combine. Once combined in the membrane the halide ion will exchange for another
ion of like or closely related species on either side of the membrane, but when chloride ions are being absorbed by the cell,
what makes them dissociate from this compound on the inner
side of the plasma membrane?

To explain these data, or indeed any anion pump occurring
in parts of the cell other than the mitochondria (or chloroplasts) themselves, we must postulate either that the energy
is transferred in the form of ATP or reducing equivalents
(e.g. NADH) or that H^+ and OH^- have been transferred, still
separated, from the organelle where the charge separation
occurred in the first place. Is this likely? We have seen that the
separation of hydrogen ion and hydroxyl ion (which becomes
bicarbonate) in the gastric mucosa is so complete that they pass
out on different sides of the cell. The present postulate is:

> That associated with membranes in the endoplasmic reti-
> culum, probably forming vesicles, separation of hydrogen
> ions (with an appropriate anion) and hydroxyl or bicarbon-
> ate ions (with an appropriate cation) is maintained. Both
> hydrogen ions and hydroxyl ions can be used by the cell

to move other ions and energy is conserved until they come together again in the same phase and water is formed.

We also know that mitochondria are capable of rhythmic swelling and contraction; in the swelling phase an adjacent vesicle would be filled with the anions pumped out of the mitochondrion and with the accompanying hydrogen ions. During the shrinkage phase, the hydroxyl ions and accompanying cations leak from the mitochondrion. If the swelling and shrinkage activity occurred in the same phase of the endoplasmic reticulum, water would be formed and no energy would be conserved. If the two successive stages of the cycle occur when the organelle is in contact with different vesicles, then the separation of hydrogen and hydroxyl ions will persist. To explain the observations of Cram it is further postulated that the chloride leaves the site of its exchange diffusion when replaced by a hydroxyl ion which then passes to the exterior as the next chloride ion from outside is absorbed. Meanwhile at other sites in the cytoplasm, vesicles filled with hydrogen ions and an accompanying anion make their way to the cell surface and there a hydrogen ion exchanges for a cation. Water would be formed at the cell surface and the net result would be an uptake of both ions of the salt.

The following evidence is in accord with this hypothesis:

(1) Mitochondria are seen to be moved rapidly around the cell in the cytoplasmic streaming. If they are undergoing successive cycles of hydrogen-ion extrusion and hydroxyl leakage, they will do so in contact with different parts of the cytoplasm.

(2) Electron-microscope evidence suggests that all plant cells so far investigated which are engaged in active salt transport contain a large number of vesicles in the cytoplasm. This is particularly well illustrated by the secretory cells of the leaves of saltbush, which accumulate sodium chloride in their vacuoles and contain many more vesicles than the adjoining (non-secretory) cells (plate 1).

(3) The evidence that the transfer of energy for the salt accumulation process in these plant cells is not in the form of ATP. Atkinson, Eckermann, Grant and Robertson (1966) showed that the ATP content of carrot tissue dropped when active transport began. This might mean no more than that the ATP reserves were being used to bring about active transport

73

but it could also mean (as the basic hypothesis suggests) that ATP was not being formed when the charge separation was used to exchange for the ions in active transport. The second piece of evidence was more definitive: oligomycin reduced the ATP content of salt-accumulating carrot tissue in a few minutes but did not slow down the accumulation of salt until about $\frac{1}{2}$ h later. These results were interpreted as showing that the salt-accumulation mechanism depended on the charge separation but not on ATP formation.

The properties of the membranes of endoplasmic reticulum are largely unknown. It is known that, apart from the protein synthesising systems associated with the ribosomes at the 'rough' endoplasmic reticulum membranes, some enzymes occur in the membranes of the smooth endoplasmic reticulum. The sodium/potassium ATP-ase (to be discussed later), various acid phosphatases and a non-phosphorylating NAD-cytochrome reductase are known to occur (Parsons, 1966). The function of the NAD-cytochrome reductase is not understood but it should be kept in mind as a possible proton-moving system in accordance with the Mitchell hypothesis; it might be an alternative method of separating hydrogen ions and hydroxyl ions if energy were transferred from the mitochondrion as reducing power.

(d) Anion pumps in animal cells

Active transport of chloride has been established in rabbit cornea (Green, 1965), in the intestine of the marine teleost (House and Green, 1965), in the gall bladder of several species (Diamond, 1962). Long ago Krogh (1937) showed than an anion pump probably accounted for the movement of chloride in goldfish gills; a similar pump has recently been demonstrated in turtle bladder (Brodsky and Schilb, 1966). The mechanisms of anion pumps in animal cells are no better understood than those of plant cells and need thorough investigation to see how far the hypotheses here proposed may be applicable.

Cation pumps

(a) The sodium pump

By far the most investigated and best understood mechanism of active transport is that of the sodium/potassium pump. A wide variety of cells has a mechanism which can take up potassium

74

and extrude sodium but it is necessary to recognize that differences in the mechanisms probably exist. Characteristically the sodium/potassium pump requires K^+ in the external medium and is inhibited by cardiac glycosides (of which ouabain is most commonly used), but there is evidence for an ouabain-insensitive sodium pump in red cells (Hoffman and Krezenow, 1966) and in many plant cells.

TABLE 5. *Sodium and potassium and potential differences in giant algal cells*

Cell	Concentration (m-equiv./l.)		Potential difference (mV) between outside and vacuole	Reference
	Medium	Vacuole		
Nitella translucens	K 0·1 Na 1·0	75 65 }	−120	MacRobbie (1962)
N. flexilis	K 0·1 Na 0·2	80 27 }	−160	Kishimoto, Nagai and Tazawa (1965) Kishimoto and Tazawa (1965 *a*)
Hydrodictyon africanum	K 0·1 Na 1·0	40 17 }	−90	Raven (1967)
Lamprothamnium succinctum	K 7 Na 300	250 140 }	−100	Kishimoto and Tazawa (1965 *b*)
Valonia ventricosa	K 12 Na 508	625 44 }	+17	Gutknecht (1966)

In plants, sodium/potassium pumps have been established for a number of the giant algal cells. The distribution of potential differences and the concentrations of sodium and potassium in some examples are shown in table 5. The high concentrations of both potassium and sodium in the vacuole, compared with outside, indicate that active transport has occurred in a number of cells, but the ratio K/Na indicates that there has been preferential uptake of potassium with exclusion of sodium and, in the *Valonia* and *Lamprothamnium* in particular, that the sodium has been actively excluded. In *Nitella translucens*, MacRobbie (1966 *a, b*) has shown that the active influx of potassium and active efflux of sodium occur probably at the plasmalemma and that the potassium uptake is dependent on ATP formation, being linked with the ATP production in proximity to light-reaction I in photosynthesis. In the cells of higher plants, Pitman (1965) suggests that a sodium/potassium pump must be operating in barley roots. Pitman and Saddler (1967) point out that there is a potassium influx pump and a sodium efflux pump operating over the low range of concentration (up to 0·2 mM) in barley roots; this conclusion is based on the two observations that (1) in the cells of the cortex the ratio K^+/Na^+ is about 3/1 when the solution had a ratio K^+/Na^+ of 1/3; (2) uptake of tracer sodium is quite high (1·0 μ-equiv*./ g/h) when the net uptake was low (only 0·02 μ-equiv./g/h), so the difference in ratio cannot be due to low permeability to sodium. Further, using their technique for discriminating between compartments in the cell, they concluded that K^+/Na^+ in the cytoplasm was about 11 compared with the 0·33 in the external solution. Taking all their results into account, the simplest explanation would be a sodium/potassium pump located in the surface of the cytoplasm or at the plasmalemma. Cram (1967) has recently found a sodium extrusion pump in carrot tissue which is inhibited by ouabain.

The sodium pumps in animal tissue are so well known and so many of their characteristics have been investigated in detail that no attempt will be made here to parade all the evidence. The subject has been well reviewed recently by Baker (1966), Albers (1967) and by Charnock and Opit (1968). One of the most important advances in study of this type of active transport came from the important observation by Skou (1957) that it was possible to isolate a membrane-bound ATP-hydrolysing enzyme from the leg nerves of a crab. This ATP-hydrolysing enzyme required the simultaneous presence of sodium and potassium as well as magnesium to exert its full hydrolytic rate. This important sodium/potassium activated ATP-ase was shown to be associated with the membranes of the smooth endoplasmic reticulum (Charnock and Post, 1963). Meanwhile, parallel evidence from mammalian erythrocytes confirmed that the sodium extrusion of those cells was dependent on ATP (there produced by glycolysis) (Whittam and Ager, 1965). In structures accumulating potassium and extruding sodium, such as the red-cell membrane, the activating ions need to be on opposite sides of the membrane. The ATP-ase is orientated across the membrane, as shown by the erythrocyte ghosts which do not hydrolyse external ATP (Glynn, 1962;

Whittam, 1962), and by the maximum hydrolysis of internal ATP with high external K^+ and high internal Na^+. The external site can be activated by K^+, Rb^+ and NH_4^+ about equally but the internal site is completely Na^+ specific. The sodium ATP-ase incubated with ATP which has ^{32}P in the γ-position incorporates the ^{32}P with the enzyme protein which can be precipitated and simultaneously liberates the ADP. This has led to the hypothesis that the reaction sequence is

$$\text{enzyme} + \text{AT}^{32}\text{P} \xrightarrow{\text{Na}} \text{enzyme}^{32}\text{P} + \text{ADP},$$

$$\text{enzyme}^{32}\text{P} + H_2O \xrightarrow{K^+} \text{enzyme} + {}^{32}\text{P}.$$

Various models have been suggested to account for the way in which this enzyme located in a membrane brings about the movement of ions simultaneous with the hydrolysis of ATP, but in reaching a conclusive hypothesis, at present we are handicapped by our lack of knowledge of the organization of these membranes. One of the most interesting models is that developed by Opit and Charnock (1965) and modified recently to take account of most recent knowledge of the system and the principles which must be involved. Whatever the details, it seems likely that, with the dimensions of the enzyme relative to the dimensions of the membrane, allosteric changes may be brought about by the interaction of protein ions and substrate, and these changes result in movement of ions from a site of combination on one side to a site of liberation on the other.

One very important property of the sodium pump has been demonstrated by Garrahan and Glynn (1966); they showed that AT^{32}P could be formed by the intact membrane of the red cell in the presence of ^{32}P and of sodium and potassium, thus showing reversal of the normal sequence of this ATP-ase to result in ATP synthesis. This reaction of the ATP-ase is clearly analogous to the reversibility of the ATP-ases of mitochondria. It is another example of the way in which energy conserved by an ion separation, here sodium and potassium, can be brought back into circulation as ATP.

(b) Other cation pumps

Apart from some evidence that the calcium might be moved specifically in the sarcoplasm of skeletal muscle, there is not much evidence for specific cation pumps (other than the sodium pump) in either plant or animal cells except in the mitochondrial membranes already discussed. The mechanism is still obscure (see Albers, 1967).

Problems of active transport in cells

This chapter has emphasized the difficulties in connecting the source of the energy (mitochondrial, chloroplast or glycolysis) with the site of the pump which must be in a membrane. It is suggested that there are only three ways in which the energy is known to be transferred: (a) as ATP, (b) as reducing power (e.g. NADH), or (c) as transfer of the H^+ or OH^- after separation. We have seen (1) that the sodium pump is undoubtedly due to the transfer of ATP; (2) that it is unlikely that the secretion of hydrochloric acid is due to anything but the separation of charge, probably occurring at the mitochondrial membrane; (3) that at present no active transport system is known to have the energy transferred in reducing equivalents. It is suggested too that the cell contains membranes which, though very thin, are very efficient in keeping ions separated except at the point of transport or exchange diffusion; these are exemplified by the endoplasmic membranes which are concerned in such mechanisms as the sodium pump and the secretion of hydrochloric acid by the mucosa. Finally, it is suggested that the complex system of membranes and vesicles, seen with the electron microscope in secretory cells of animals and in accumulating cells of plants, is concerned with the control of the ionic distribution and the position of exchange diffusion in the cytoplasm.

CHAPTER 6

The Next Approximation

'The scientist does not speak of the last analysis but rather of the next approximation.' (G. N. LEWIS)

THE separation of protons and electrons at living membranes is a very important fundamental biological phenomenon. We have long known that, starting with water, the photosynthetic process separates hydrogen atoms from oxygen and ultimately this reducing power results in the bond energy in carbohydrates which becomes the source of all energy for living organisms. We have recognized too that the energy is transferred from one part of the cell to another in phosphate-bond energy with the ubiquitous carrier, ATP, responsible. We have recognized that the formation of ATP can occur either directly in photosynthesis or, more generally, as the result of phosphorylation at the level of certain substrate oxidations (as in glycolysis) or in the electron-transport system of oxidative phosphorylation. We have perhaps been slow to recognize that cells with a mechanism for separating protons and electrons or hydrogen and hydroxyl ions would have a powerful source of energy as long as these charges remain apart. Once they react again the energy is no longer conserved, as the equilibrium lies so definitely in the direction of water formation. As pointed out previously (Robertson, 1960) the energy conserved by charge separation in the respiratory process can be compared in the two equations

$$C_6H_{12}O_6 + 6H_2O + 6O_2 \longrightarrow 6CO_2 + 12H_2O + 680\,000 \text{ cal}$$

and

$$C_6H_{12}O_6 + 18H_2O + 6O_2$$
$$\longrightarrow 6CO_2 + 24H^+ + 24OH^- + 458\,000 \text{ cal,}$$

that is, about one-third of the energy liberated by the reaction would be conserved as charge separation. We have discussed the evidence that in both photophosphorylation and oxidative phosphorylation the charge separation precedes the phosphorylation of ADP to give ATP, but it is important to regard the processes involved as potentially reversible. The work on chloroplasts and mitochondria clearly illustrates the phosphorylation resulting from H^+ and OH^- across an appropriate membrane with the proper enzyme. The processes in these organelles are summarized in fig. 16. These are the reactions

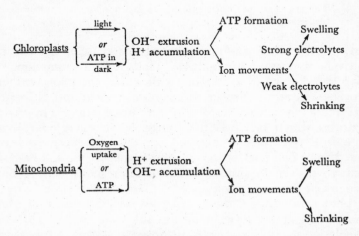

Fig. 16. Summary of the effects of charge separation on associated processes in chloroplasts and mitochondria.

to be expected in organelles but we must consider two other important possibilities:

(1) There is no theoretical reason why the ion movements resulting from H^+ and OH^- movements in exchange should not also be reversible, though this has not yet been demonstrated; if this does turn out to be possible, it would mean that energy conserved in the active transport of ions (e.g. of a salt like KCl in plant cells) can be re-used in the cellvia an exchange for OH^- and H^+ ions. H^+ and OH^- must be kept apart and transferred to the reversible membrane-bound ATP-ase, such as found in either chloroplasts or mitochondria, where ATP would be formed. While this reversibility has not yet been demonstrated experimentally, it would be exactly analogous to the reversal of the sodium–potassium ATP-ase to

80

synthesize ATP as demonstrated by Garrahan and Glynn (1966). The hypothetical systems are compared in fig. 17. These interesting relationships suggest that cells may be able to draw on the energy which has been conserved as an ion

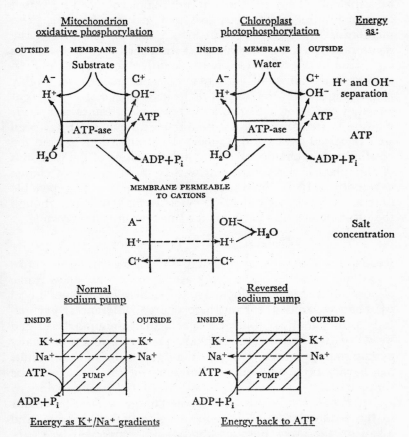

Fig. 17. Hypothetical systems for charge separation, phosphorylation and ion movement.

gradient and reconvert it to ATP—the universal energy currency of the cell—from which it may be transferred to any other cellular processes.

(2) The membranes are responsible for the separation in the first place, and for maintenance of the separated ions later and are very efficient in both processes. Despite the large

amount of work on living membranes we are still very ignorant of their properties and structure at the molecular level which would explain these phenomena. It is becoming increasingly clear that a very wide difference of ion composition and a very big range of pH can exist in adjacent regions of a cell separated by a membrane only about 50Å in thickness. Further, these membranes have distinct 'sidedness'; that is, one side of the membrane has different properties from the other and remains organized in this way. Despite these differences the membrane is altered dynamically with reacting substances, presumably interacting with the membrane-sited enzymes. Different enzymes are known to occur in these membranes and their relation to structure will not be discussed further. It has been well reviewed by Parsons (1966). It is important to note the possibility that charge separation might be one of the functions of endoplasmic membranes, as well as maintenance of charge differences. It is interesting to speculate on this possible function for microsomal NAD-cytochrome reductase, though the ultimate electron acceptor for this system is unknown.

Consequences of charge separation

Realization that such large charge differences can occur across such short distances in the cytoplasm opens a whole new range of possibilities for physiological and biochemical action. Local pH changes, consequent on these charge differences, particularly as they may approach the spectacular amounts illustrated by the gastric mucosa, would have profound effects on many cytoplasmic processes. The possible significance of these results has hardly been realized yet but some of the effects can be summarized.

(1) *Potential differences.* Where the charge separation results in the accumulation of acid on one side of a membrane and alkali on the other side, a potential difference will be established and, given an electrode system connecting the two, current can be drawn.

(2) *Energy conservation.* As long as the H^+ and OH^- ions are separated, energy will be conserved for the cell and, as we have seen, this can be made available again as ATP in the effect on the reversible ATP-ase.

(3) *Ion movements.* The exchange of the H^+ or OH^- can be used in the cell to move cations and anions respectively.

(4) *Effects on the weak electrolytes of the cell.* The liberation of acid or alkali from organelles or from the endoplasmic vesicles will have marked effects on the ionization of weak electrolytes. For example, the build-up of acid under conditions when a mitochondrion was extruding H^+ ions in quantity would suppress the ionization of weak electrolytes in the vicinity and this, in turn, could completely alter the enzyme activity, the permeability or the structural conformity of that part of the cell. We have seen how these pH changes can influence both the permeability and the structure of plastid lamellae.

(5) *Energy reactions in the cell.* These reactions, in which electron transfer is involved, would be influenced directly if the electrons were passed from the charge-separating membrane direct to the reacting intermediates. We are familiar with this need in the electron-transport chain of respiration and photosynthesis themselves. We now have to consider how far such an electron passage can affect other processes, such as the fixation of atmospheric nitrogen where the six electrons necessary to reduce an N_2 molecule to two NH_3 molecules might be transferred by such a membrane-associated, charge-separating process. Further, the Mitchell hypothesis of charge separation might be the clue to why the electron transport system has molecules both of the soluble NAD type and the insoluble cytochrome type.

(6) *Molecules transported.* There is evidence that some molecules are transported through cell membranes by first being phosphorylated and the association of glucose transport with cations is well established. We have the possibility that it is because of the charge on the molecule that the phosphorylated sugar is transported by exchange diffusion, indirectly dependent on the charge separation of H^+ and OH^-.

It is too early to say how far these effects may be related to the spectacular but inadequately explained phenomenon of protoplasmic streaming or cyclosis in plants, but undoubtedly potential differences which could be generated by charge separation in cytoplasmic membranes could be important. One mystery of plant physiology may find an explanation, as suggested by Siegenthaler and Packer (1965), though their explanation may need modification: it has been known for many years that a change in pH of guard cells in plants on illumination accompanies the change of starch in the cell to the sugar which causes an increase in osmotic pressure, draws in water, causes the guard cell to swell and hence to open the stomate. It may be that the pH change on illumination is similar to that observed in isolated chloroplast fragments and that this pH change may be responsible for activating the enzymes concerned in the starch–sugar balance.

Future developments

Some problems of fundamental interest await the development of suitable techniques for their investigation. It seems, for example, that profound changes of pH and ionic composition probably occur in vesicles which are too small to be investigated with the ordinary microscope. New techniques for studying the changes in properties in these vesicles with electron microscopy should be rewarding, and perhaps here more use can be made of tritiated compounds and the fine-grained emulsions now being applied in electron microscopy (Salpeter, 1967).

Not the least interesting development will be in the study of how far these reactions and processes in the living cell are reversible; in fact, reversibility is perhaps the most characteristic feature common to chemical reactions in living organisms. The two spectacularly reversible processes dependent on an ion separation and a charge separation respectively in ATP synthesis from Na/K unbalance (Garrahan and Glynn, 1966) and chlorophyll fluorescence from pH difference (Mayne and Clayton, 1966) may be pointers to others.

Perhaps the most interesting problem lies in the details of the charge-separating mechanisms and their relation to membrane structure. Here the newer techniques of physical chemistry applied to biological systems can be expected to yield interesting information.

Investigations of this type and others beyond my imagination will before long move us from the 'next approximation' to the next-but-one approximation.

REFERENCES

ALBERS, R. W. (1967). Biochemical aspects of active transport. *A. Rev. Biochem.* **36**, 727–56.

AMOORE, J. E. (1960). Exchange of potassium ions across a concentration difference by isolated rat-liver mitochondria. *Biochem. J.* **76**, 438–44.

AMOORE, J. E. and BARTLEY, W. (1958). The permeability of isolated rat-liver mitochondria to sucrose, sodium chloride and potassium chloride at 0 °C. *Biochem. J.* **69**, 223–36.

ARISZ, W. H. and SOL, H. H. (1956). Influence of light and sucrose on the uptake and transport of chloride in *Vallisneria* leaves. *Acta bot. neerl.* **5**, 218–38.

ATKINSON, M. R., ECKERMANN, G., GRANT, M. and ROBERTSON, R. N. (1966). Salt accumulation and adenosine triphosphate in carrot xylem tissue. *Proc. natn. Acad. Sci. U.S.A.* **55**, 560–4.

AZZI, A., ROSSI, E. and AZZONE, G. F. (1966). Swelling and shrinkage phenomena in liver mitochondria. V. Permeability of the mitochondrial membrane to univalent cations: effect of Mg^{++}. *Enzym. Biol. Clin.* **7**, 25–37.

AZZONE, G. F. and AZZI, A. (1965). Volume changes induced by inorganic phosphate in liver mitochondria. *Biochem. J.* **94**, 10C–11C.

BAKER, P. F. (1966). The sodium pump. *Endeavour* **25**, 166–72.

BARTLEY, W. and DAVIES, R. E. (1952). Secretory activity of mitochondria. *Biochem. J.* **52**, xx.

BARTLEY, W. and DAVIES, R. E. (1954). Active transport of ions by subcellular particles. *Biochem. J.* **57**, 37–49.

BENNET-CLARK, T. A. (1956). Salt accumulation and mode of action of auxin; a preliminary hypothesis. In *The Chemistry and Mode of Action of Plant Growth Substances*, ed. R. L. Wain and F. Wightman. London: Butterworth.

BENNETT, H. S. (1956). The concept of membrane flow and membrane vesiculation as mechanisms for active transport and ion pumping. *J. Biophys. Biochem. Cytol.* **2**, 94–103.

BOERI, E. and TOSI, L. (1954). The effect of salts on the autoxidation of cytochrome *c*. *Archs. Biochim. Biophys.* **53**, 83–92.

BOVÉ, J. M., BOVÉ, C., WHATLEY, F. R. and ARNON, D. I. (1963). Chloride requirement for oxygen evolution in photosynthesis. *Z. Naturforsch.* **18***b*, 683–8.

BOWLING, D. J. F., MACKLON, A. E. S. and SPANSWICK, R. M. (1966). Active and passive transport of the major nutrient ions across the root of *Ricinus communis*. *J. exp. Bot.* **17**, 410–16.

BRIERLEY, G. P., BACHMANN, E. and GREEN, D. E. (1962). Active transport of inorganic phosphate and magnesium ions by beef heart mitochondria. *Proc. natn. Acad. Sci. U.S.A.* **48**, 1928–35.

BRIERLEY, G., MURER, E., BACHMANN, E. and GREEN, D. E. (1963). Studies on ion transport. II. The accumulation of inorganic phosphate and magnesium ions by heart mitochrondira. *J. biol. Chem.* **238**, 3482-9.

BRIGGS, G. E. (1930). The accumulation of electrolytes in plant cells—a suggested mechanism. *Proc. R. Soc. B.* **107**, 248-69.

BRIGGS, G. E., HOPE, A. B. and ROBERTSON, R. N. (1961). *Electrolytes and Plant Cells.* Oxford: Blackwell.

BRODSKY, W. A. and SCHILB, T. P. (1966). Ionic mechanisms for sodium and chloride transport across turtle bladders. *Am. J. Physiol.* **210**, 987-96.

BROOKS, S. C. (1929). The accumulation of ions in living tissue—a nonequilibrium condition. *Protoplasma.* **8**, 389-412.

CALDWELL, P. C. (1956). The effects of certain metabolic inhibitors on the phosphate esters of the squid giant axon. *J. Physiol., Lond.* **132**, 35-6P.

CALDWELL, P. C. and KEYNES, R. D. (1957). The utilization of phosphate bond energy for sodium extrusion from giant axons. *J. Physiol., Lond.* **137**, 12-13P.

CARAFOLI, E., ROSSI, C. S. and LEHNINGER, A. L. (1964). Cation and anion balance during active accumulation of Ca^{++} and Mg^{++} by isolated mitochondria. *J. biol. Chem.* **239**, 3055-61.

CHANCE, B. (1964). Controlled states in mitochondria—state 6. *Fedn. Proc.* **23**, 265.

CHANCE, B. (1965). The energy-linked reaction of calcium with mitochondria. *J. biol. Chem.* **240**, 2729-48.

CHANCE, B. and MELA, L. (1966a). A hydrogen ion concentration gradient in a mitochondrial membrane. *Nature, Lond.* **212**, 369-72.

CHANCE, B. and MELA, L. (1966b). Proton movements in mitochondrial membranes. *Nature, Lond.* **212**, 372-6.

CHAPPELL, J. B., COHN, M. and GREVILLE, G. D. (1963). The accumulation of divalent ions by isolated mitochondria. In *Energy-linked Functions of Mitochondria*, pp. 219-31. New York: Academic Press.

CHAPPELL, J. B. and CROFTS, A. R. (1965). Gramicidin and ion transport in isolated liver mitochondria. *Biochem. J.* **95**, 393-402.

CHAPPEL, J. B. and CROFTS, A. R. (1966). Ion transport and reversible volume changes of isolated mitochondria. In *Regulation of Metabolic Processes in Mitochondria*, ed. J. M. Tager, S. Papa, E. Quagliariello and E. C. Slater. Amsterdam: Elsevier Publishing Co.

CHAPPELL, J. B. and GREVILLE, G. D. (1963). The influence of the composition of the suspending medium on the properties of mitochondria. *Biochem. Soc. Symp.* **23**, 39-65.

CHAPPELL, J. B. and HAARHOFF, K. N. (1967). The penetration of mitochondrial membrane by anions and cations. In *Biochemistry of Mitochondria*, ed. E. C. Slater, Z. Kaninga and L. Wojtczak. London and New York: Academic Press.

CHAPPELL, J. B., HENDERSON, P. J. F., McGIVAN, J. D. and ROBINSON, B. H. (1967). The effect of drugs on mitochondrial functions. In *Symposium on the Interaction of Drugs and Sub-cellular Components in Animal Tissue*, ed. P. N. Campbell. London: J. and A. Churchill.

86

CHARNOCK, J. S. and OPIT, L. J. (1967). Membrane metabolism and ion transport. In *The Biological Basis of Medicine*, Vol. I, pp. 69–103. Ed. E. E. Bitter, London: Academic Press.

CHARNOCK, J. S. and POST, R. L. (1963). Studies on the mechanism of cation transport. I. The preparation and properties of a cation-stimulated adenine triphosphatase from guinea-pig kidney cortex. *Aust. J. exp. Biol. med. Sci.* **41**, 547–60.

COCKRELL, R. S., HARRIS, E. J. and PRESSMAN, B. C. (1966). Energetics of potassium transport in mitochondria induced by valinomycin. *Biochem.* **5**, 2326–35.

CONWAY, E. J. (1959). The redox pump theory and the present evidence. In *The Method of Isotopic Tracers Applied to the Study of Active Transport*. London: Pergamon Press.

CONWAY, E. J. and BRADY, T. C. (1948). Source of the hydrogen ions in gastric juice. *Nature, Lond.* **162**, 456–7.

CRAM, W. J. (1967). Compartmentation and control of inorganic ions in higher plant cells. Ph.D. Dissertation, University of Cambridge.

CRANE, E. E. and DAVIES, R. E. (1948*a*). Chemical energy relations in gastric mucosa. *Biochem. J.* **43**, xlii.

CRANE, E. E. and DAVIES, R. E. (1948*b*). Electric energy relations in gastric mucosae. *Biochem. J.* **43**, xlii.

CRANE, E. E. and DAVIES, R. E. (1951). Chemical relations in gastric mucosae. *Biochem. J.* **49**, 169–75.

DAVENPORT, H. W. (1957). Metabolic aspects of gastric acid secretion. In *Metabolic Aspects of Transport Across Cell Membranes*, pp. 295–302. Madison: University of Wisconsin Press.

DAVIES, R. E. (1951). The mechanism of hydrochloric acid production by the stomach. *Biol. Rev.* **26**, 87–120.

DAVIES, R. E. (1957). Gastric hydrochloric acid production—the present position. In *Metabolic Aspects of Transport Across Cell Membranes*, pp. 277–93. Madison: University of Wisconsin Press.

DAVIES, R. E. (1961). Discussion in *Membrane Transport and Metabolism*, pp. 320–3, ed. A. Kleinzeller and A. Kotyk. New York: Academic Press.

DAVIES, R. E. and KREBS, H. A. (1952). Biochemical aspects of the transport of ions by nervous tissue. *Symp. Biochem. Soc.* **8**, 77–92.

DAVIES, R. E. and OGSTON, A. G. (1950). On the mechanism of secretion of ions by gastric mucosae and by other tissues. *Biochem. J.* **46**, 324–33.

DELUCA, H. F. and ENGSTRÖM, G. W. (1961). Calcium uptake by rat kidney mitochondria. *Proc. natn. Acad. Sci. U.S.A.* **47**, 1744–50.

DELUCA, H. F., ENGSTRÖM, G. W. and RASMUSSEN, H. (1962). The action of vitamin D and parathyroid hormone *in vitro* on calcium uptake and release by kidney mitochondria. *Proc. natn. Acad. Sci. U.S.A.* **48**, 1604–9.

DIAMOND, J. M. (1962). The mechanism of solute transport by the gall bladder. *J. Physiol., Lond.* **161**, 474–502.

DILLEY, R. A. (1967). Ion and water transport processes in spinach chloroplasts. *Brookhaven Symp. Biol.* **19**, 258–79.

DILLEY, R. A. and VERNON, L. P. (1965). Ion and water transport processes related to the light dependent shrinkage of spinach chloroplasts. *Archs. Biochem. Biophys.* **111**, 365–75.

Du Buy, H. C., Woods, M. W. and Lackey, M. D. (1950). Enzymatic activities of isolated, normal and mutant mitochondria and plastids of higher plants. *Science, N.Y.* **111**, 572–4.

Epstein, E. (1966). Dual pattern of ion absorption by plant cells and by plants. *Nature, Lond.* **212**, 1324–7.

Epstein, E. and Hagen, C. E. (1952). A kinetic study of the absorption of alkali cations by barley roots. *Pl. Physiol.* **27**, 457–74.

Epstein, E., Rains, D. W. and Elzam, O. Z. (1963). Resolution of dual mechanisms of potassium absorption in barley roots. *Proc. natn. Acad. Sci. U.S.A.* **49**, 684–92.

Farrant, J. L., Potter, C., Robertson, R. N. and Wilkins, M. J. (1956). The structure of plant mitochondria. *Aust. J. Bot.* **4**, 117–24.

Farrant, J. L., Robertson, R. N. and Wilkins M. J. (1953). The mitochondrial membrane. *Nature, Lond.* **171**, 401–2.

Friedenwald, J. S. and Stiehler, R. D. (1938). A mechanism of circulation of the aqueous secretion of the intraocular fluid. *Arch. Ophthal. N.Y.* **20**, 761–86.

Garrahan, P. J. and Glynn, I. M. (1966). Driving the sodium pump backwards to form adenosine triphosphate. *Nature, Lond.* **211**, 1414–15.

Glynn, I. M. (1962). Activation of adenosine triphosphatase activity in a cell membrane by external potassium and internal sodium. *J. Physiol., Lond.* **160**, 18P.

Goldacre, R. J. (1952). The folding and unfolding of protein molecules as a basis of osmotic work. *Int. Rev. Cytol.* **1**, 135.

Gray, J. S. (1943). The physiology of the parietal cell with special reference to the formation of acid. *Gasteroenterology*, **1**, 390–400.

Green, K. J. (1965). Ion transport in isolated cornea of the rabbit. *Am. J. Physiol.* **209**, 1311–16.

Greenham, C. G. (1966). The relative electrical resistances of the plasmalemma and the tonoplast in higher plants. *Planta.* **69**, 150–7.

Gutknecht, J. (1966). Sodium, potassium and chloride transport and membrane potentials in *Valonia ventricosa*. *Biol. Bull. mar. biol. Lab., Woods Hole*, **130**, 331–44.

Hackenbrock, C. R. (1966). Ultrastructural bases for metabolically linked mechanical activity in mitochondria. I. Reversible ultrastructural changes with change in metabolic steady state in isolated liver mitochondria. *J. Cell Biol.* **30**, 269–97.

Higinbotham, N., Etherton, B. and Foster, R. J. (1967). Mineral ion contents and cell transmembrane electropotentials of pea and oat seedling tissue. *Pl. Physiol.* **42**, 37–46.

Hill, R. and Bendall, F. (1960). Function of two cytochrome components in chloroplasts: a working hypothesis. *Nature, Lond.* **186**, 136–7.

Hind, G. and Jagendorf, A. T. (1963). Separation of light and dark stages in photophosphorylation. *Proc. natn. Acad. Sci. U.S.A.* **49**, 715–22.

Hind, G. and Jagendorf, A. T. (1965). Light scattering changes associated with the production of a possible intermediate in photophosphorylation. *J. biol. Chem.* **240**, 3195–201.

HOFFMAN, J. F. and KREZENOW, F. M. (1966). The characterization of new energy dependent cation transport processes in red blood cells. *Ann. N.Y. Acad. Sci.* **137**, 566–76.

HOGBEN, C. A. M. (1951). The chloride transport system of the gastric mucosa. *Proc. natn. Acad. Sci. U.S.A.* **37**, 393–5.

HOGBEN, C. A. M. (1955). Active transport of chloride by isolated frog gastric epithelium: origin of the gastric mucosal potential. *Am. J. Physiol.* **180**, 641–9.

HONDA, S. I. and ROBERTSON, R. N. (1956). Studies in the metabolism of plant cells. XI. The Donnan equilibration and ionic relations of plant mitochondria. *Aust. J. Biol. Sci.* **9**, 305–20.

HONDA, S. I., ROBERTSON, R. N. and GREGORY, J. M. (1958). Studies in the metabolism of plant cells. XII. Ionic effects on oxidation of reduced diphosphate-pyridine nucleotide and cytochrome *c* by plant mitochondria. *Aust. J. Biol. Sci.* **11**, 1–15.

HOUSE, C. R. and GREEN, K. J. (1965). Ions and water transport in isolated intestine of the marine teleost, *Cottus scorpius. J. exp. Biol.* **42**, 177–89.

IZAWA, S. and GOOD, N. E. (1966). Effect of salts and electron transport on the conformation of isolated chloroplasts. I. Light-scattering and volume changes. *Pl. Physiol.* **41**, 533–43.

IZAWA, S. and HIND, G. (1967). The kinetics of pH rise in illuminated suspensions. *Biochim. biophys. Acta.* **143**, 377–90.

JAGENDORF, A. T. and HIND, G. (1963). Studies on the mechanism of photophosphorylation. In *Photosynthetic Mechanisms of Green Plants*, pp. 599–610. Washington: Nat. Acad. Sci.—Nat. Res. Council Publ. no. 1145.

JAGENDORF, A. T. and URIBE, E. (1966). ATP formation caused by acid base transition of spinach chloroplasts. *Proc. natn. Acad. Sci. U.S.A.* **55**, 170–7.

JAGENDORF, A. T. and URIBE, E. (1967). Photophosphorylation and the chemi-osmotic hypothesis. In *Energy Conversion by the Photosynthetic Apparatus*, pp. 215–41. *Brookhaven Symp. Biol.*, no. 19.

KISHIMOTO, U. and TAZAWA, M. (1965a). Ionic composition of the cytoplasm of *Nitella flexilis. Pl. cell Physiol., Tokyo*, **6**, 507–18.

KISHIMOTO, U. and TAZAWA, M. (1965b). Ionic composition and electric response of *Lamprothamnium succinctum. Pl. Cell Phys.* **6**, 529–36.

KISHIMOTO, U., NAGAI, R. and TAZAWA, M. (1965). Plasmalemma potential in *Nitella. Pl. Cell Physiol.* **6**, 519–28.

KREZENOW, F. M. and HOFFMAN, J. F. (1966). The characterization of new energy dependent cation transport processes in red blood cells. *Ann. N.Y. Acad. Sci.* **137**, 566–76.

KROGH, A. (1937). Osmotic regulation in fresh water fishes by active absorption of chloride ions. *Z. Vergl. Physiol.* **24**, 656–66.

LEHNINGER, A. L. (1964). *The Mitochondrion.* New York: W. A. Benjamin Inc.

LINNANE, A. W. (1958). A soluble component required for oxidative phosphorylation by a sub-mitochondrial particle from beef heart muscle. *Biochim. biophys. Acta.* **30**, 221–2.

89

LOOMIS, W. F. and LIPMAN, F. (1948). Reversible inhibition of the coupling between phosphorylation and oxidation. *J. biol. Chem.* **173**, 807–8.

LUND, E. J. (1928). Relation between continuous bio-electric currents and cell respiration. *J. exp. Zool.* **51**, 265–307.

LUNDEGÅRDH, H. (1939). An electro-chemical theory of salt absorption and respiration. *Nature, Lond.* **143**, 203.

LUNDEGÅRDH, H. (1940). Investigations as to the absorption and accumulation of inorganic ions. *Ann. Agr. Col. Sweden.* **8**, 234–404.

LUNDEGÅRDH, H. (1945). Absorption, transport and exudation of inorganic ions by the roots. *Ark. Bot.* **32 A**, 1–139.

LUNDEGÅRDH, H. (1953). Controlling effect of salts on the activity of the cytochrome oxidase. *Nature, Lond.* **171**, 477–8.

LUNDEGÅRDH, H. (1966). *Plant Physiology.* Edinburgh and London: Oliver and Boyd.

LUNDEGÅRDH, H. and BURSTRÖM, H. (1933). Untersuchungen über die Salzaufnahme der Pflanzen. III. Quantitative Beziehungen zwischen Atmung und Anionaufnahme. *Biochem. Z.* **261**, 235–51.

LUNDEGÅRDH, H. and BURSTRÖM, H. (1935). Untersuchungen über die Atmungsvorgänge in Pflanzenwurzeln. *Biochem. Z.* **277**, 223–49.

McCARTY, R. E. and RACKER, E. (1966a). Effects of an antiserum to the chloroplast coupling factor on photophosphorylation and related processes. *Fedn. Proc.* **25**, 226.

McCARTY, R. E. and RACKER, E. (1966b). Effect of a coupling factor and its antiserum on photophosphorylation and hydrogen ion transport. In *Brookhaven Symp. Biol.* **19**, 202–12.

McDONALD, I. R., BACON, J. S. D., VAUGHAN, D. and ELLIS, R. J. (1966). The relation between ion absorption and protein synthesis in beet disks. *J. exp. Bot.* **17**, 822–37.

MACFARLANE, M. G. and SPENCER, A. G. (1953). Changes in the water, sodium and potassium content of rat liver mitochondria during metabolism. *Biochem. J.* **54**, 569–75.

MACROBBIE, E. A. C. (1962). Ionic relations of *Nitella translucens*. *J. gen. Physiol.* **45**, 861–77.

MACROBBIE, E. A. C. (1964). Factors affecting the fluxes of potassium and chloride ions in *Nitella translucens*. *J. gen. Physiol.* **47**, 859–77.

MACROBBIE, E. A. C. (1965). The nature of the coupling between light energy and active ion transport in *Nitella translucens*. *Biochim. biophys. Acta.* **94**, 64–73.

MACROBBIE, E. A. C. (1966a). Metabolic effects on ion fluxes in *Nitella translucens*. I. Active influxes. *Aust. J. Biol. Sci.* **19**, 363–70.

MACROBBIE, E. A. C. (1966b). Metabolic effects on ion fluxes in *Nitella translucens*. II. Tonoplast fluxes. *Aust. J. Biol. Sci.* **19**, 371–83.

MAYNE, B. C. and CLAYTON, R. K. (1966). Luminescence of chloroplasts induced by acid-base transitions. *Proc. natn. Acad. Sci. U.S.A.* **55**, 494–7.

MERCER, F. V., HODGE, A. J., HOPE, A. B. and McLEAN, J. D. (1955). The structure and swelling properties of *Nitella* chloroplasts. *Aust. J. Biol. Sci.* **8**, 1–18.

MILLARD, D. L., WISKICH, J. T. and ROBERTSON, R. N. (1964). Ion uptake by plant mitochondria. *Proc. natn. Acad. Sci. U.S.A.* **52**, 996–1004.

MILLARD, D. L., WISKICH, J. T. and ROBERTSON, R. N. (1965). Ion uptake and phosphorylation in mitochondria; effect of monovalent ions. *Pl. Physiol.* **40**, 1129–35.

MILLER, G. W. and EVANS, H. J. (1956). The influence of salts on the activity of particulate cytochrome oxidase from roots of higher plants. *Pl. Physiol.* **31**, 357–64.

MITCHELL, P. (1961). Coupling of phosphorylation to electron and hydrogen transfer by a chemi-osmotic type of mechanism. *Nature, Lond.* **191**, 144–8.

MITCHELL, P. (1962). Molecule, group and electron translocation through natural membranes. *Biochem. Soc. Symp.* **22**, 142–69.

MITCHELL, P. (1966). Chemi-osmotic coupling in oxidative and photosynthetic phosphorylation. *Biol. Rev.* **41**, 445–602.

MITCHELL, P. and MOYLE, J. (1965). Stoichiometry of proton translocation through the respiratory chain and adenosine triphosphatase systems of rat liver mitochondria. *Nature, Lond.* **208**, 147–51.

MITCHELL, P. and MOYLE, J. (1967). Proton-transport phosphorylation: some experimental tests. In *Biochemistry of Mitochondria*, ed. E. C. Slater, Z. Kaninga and L. Wojtczak. London: Academic Press.

NEUMANN, J. S. and JAGENDORF, A. T. (1964). Light-induced pH changes related to photophosphorylation by chloroplasts. *Archs. Biochem. Biophys.* **107**, 109–19.

NOBEL, P. S. (1967). Calcium uptake, ATP-ase and photophosphorylation by chloroplasts *in vitro*. *Nature, Lond.* **214**, 875–7.

NOBEL, P. S. and MURAKAMI, S. (1967). Electron microscope evidence for the location and amount of ion accumulation by spinach chloroplasts. *J. Cell Biol.* **32**, 209–11.

NOBEL, P. S., MURAKAMI, S. and TAKAMIYA, A. (1966). Localization of light-induced strontium accumulation in spinach chloroplasts. *Pl. Cell Physiol.* **7**, 263–75.

NOBEL, P. S. and PACKER, L. (1964a). Energy-dependent ion uptake in spinach chloroplasts. *Biochim. biophys. Acta.* **88**, 453–5.

NOBEL, P. S. and PACKER, L. (1964b). Studies on ion translocation by spinach chloroplasts. *J. Cell Biol.* **23**, 67A–68A.

OPIT, L. J. and CHARNOCK, J. S. (1965). A molecular model for a sodium pump. *Nature, Lond.* **208**, 471–4.

PACKER, L. (1963). Structural changes correlated with photochemical phosphorylation in chloroplast membranes. *Biochim. biophys. Acta.* **75**, 12–22.

PACKER, L., DEAMER, D. W. and CROFTS, A. R. (1967). Conformational changes in chloroplasts. *Brookhaven Symp. Biol.* **19** [BNL 989 (C-48)], 281–302.

PACKER, L. and MARCHANT, R. M. (1964). Action of adenosine triphosphate on chloroplast structure. *J. biol. Chem.* **18**, 2061–9.

PACKER, L., SIEGENTHALER, P. A. and NOBEL, P. S. (1965). Light-induced volume changes in spinach chloroplasts. *J. Cell Biol.* **26**, 593–9.

PACKER, L., UTSUMI, K. and MUSTAFA, M. (1966). Oscillatory states of mitochondria. I. Electron and energy transfer pathways. *Arch. Biochem. Biophys.* **117**, 381-93

PARSONS, D. F. (1966). *Ultrastructural and Molecular Aspects of Cell Membranes.* 7th Canadian Cancer Research Conference, Honey Harbour, Ontario.

PATTERSON, W. B. and STETTEN, D. (1949). A study of gastric HCl formation. *Science, N.Y.* **109**, 256–8.

PFAFF, E., KLINGENBERG, M. and HELDT, H. W. (1965). Unspecific permeation and specific exchange of adenine nucleotides in liver mitochondria. *Biochem. biophys. Acta.* **104**, 312–15.

PITMAN, M. G. (1964). The effect of divalent cations on the uptake of salt by beetroot tissue. *J. exp. Biol.* **15**, 444–56.

PITMAN, M. G. (1965). Sodium and potassium uptake by seedlings of *Hordeum vulgare. Aust. J. Biol. Sci.* **18**, 10–24.

PITMAN, M. G. and SADDLER, H. D. W. (1967). Active sodium and potassium transport in cells of barley roots. *Proc. natn. Acad. Sci. U.S.A.* **57**, 44–9.

PRESSMAN, B. C. (1963). Specific inhibitors of energy transfer. In *Energy Linked Functions of Mitochondria.* New York: Academic Press.

PRESSMAN, B. C. (1965). Induced active transport of ions in mitochondria. *Proc. natn. Acad. Sci. U.S.A.* **53**, 1076–83.

PULLMAN, M. E., PENEFSKY, H. S. and RACKER, E. (1958). A soluble protein fraction required for coupling phosphorylation to oxidation in submitochondrial fragments of beef heart mitochondria. *Archs. Biochem. Biophys.* **76**, 227–30.

PULLMAN, M. E. and SCHATZ, G. (1967). Mitochondrial oxidations and energy coupling. *A. Rev. Biochem.* **36**, 539–610.

RAVEN, J. A. (1967). Ion transport in *Hydrodictyon africanum. J. gen. Physiol.* **50**, 1607–25.

REID, R. A., MOYLE, J. and MITCHELL, P. (1966). Synthesis of adenosine triphosphate by a proton motive force in rat liver mitochondria. *Nature, Lond.* **212**, 257–8.

ROBERTSON, R. N. (1951). Mechanism of absorption and transport of inorganic nutrients in plants. *A. Rev. Pl. Phys.* **2**, 1–24.

ROBERTSON, R. N. (1960). Ion transport and respiration. *Biol. Rev.* **35**, 231–64.

ROBERTSON, R. N. (1967). The separation of protons and electrons as a fundamental biological process. *Endeavour* **26**, 134–9.

ROBERTSON, R. N. and THORN, M. (1945). Studies in the metabolism of plant cells. IV. The reversibility of the salt respiration. *Aust. J. exp. Biol. Med. Sci.* **23**, 306–9.

ROBERTSON, R. N. and WILKINS, M. J. (1948a). Quantitative relation between salt accumulation and salt respiration in plant cells. *Nature, Lond.* **161**, 101.

ROBERTSON, R. N. and WILKINS. M. J. (1948b). Studies in the metabolism of plant cells. VII. The quantitative relation between salt accumulation and salt respiration. *Aust. J. Sci. Res.* B. **1**, 17–37.

ROBERTSON, R. N., WILKINS, M. J., HOPE, A. B. and NESTEL, L. (1955). Studies in the metabolism of plant cells. X. Respiratory activity and ionic relations of plant mitochondria. *Aust. J. Biol. Sci.* **8**, 164–85.

ROBERTSON, R. N., WILKINS, M. J. and WEEKS, D. C. (1951). Studies in the metabolism of plant cells. IX. The effects of 2,4-dinitrophenol on salt accumulation and salt respiration. *Aust. J. Sci. Res.* B. **4**, 248–64.

SALPETER, M. (1967). Electron microscope radioautography as a quantitative tool in enzyme cytochemistry. I. The distribution of acetylcholinesterase at motor end plates of a vertebrate twitch muscle. *J. Cell Biol.* **32**, 379–89.

SCHRÖDINGER, E. (1948). *What is Life?* Cambridge University Press.

SHEN, Y. K. and SHEN, G. M. (1962). Studies on photophosphorylation. *Scientia sin.* **11**, 1097–1106.

SIEGENTHALER, A. and PACKER, L. (1965). Light-dependent volume changes and reactions in chloroplasts. I. Action of alkenylsuccinic acids and phenylmercuric acetate and possible relation to mechanisms of stomatal control. *Pl. Physiol.* **40**, 785–91.

SKOU, J. C. (1957). The influence of some cations on an adenosine triphosphatase from peripheral nerves. *Biochim. biophys. Acta.* **23**, 394–401.

SLATER, E. C. (1966). Oxidative phosphorylation. In *Comprehensive Biochemistry*, vol. XIV, pp. 327–96. Amsterdam: Elsevier.

SLATER, E. C. (1967). An evaluation of the Mitchell hypothesis of chemiosmotic coupling in oxidative and photosynthetic phosphorylation. *European J. Biochem.* **1**, 317–26.

SMITH, F. A. (1966). Active phosphate uptake by *Nitella translucens*. *Biochim. biophys. Acta.* **126**, 94–9.

SNOSWELL, A. M. (1966). Respiration-dependent proton movements in rat liver mitochondria. *Biochem.* **5**, 1660–6.

SPANNER, D. C. (1964). *Introduction to Thermodynamics*. London and New York: Academic Press.

STANBURY, S. W. and MUDGE, G. H. (1953). Potassium metabolism of liver mitochondria. *Proc. Soc. exp. Biol. Med.* **82**, 675–81.

STEWARD, F. C. (1954). Salt accumulation in plants: a reconsideration of the role of growth and metabolism. *Symp. Soc. exp. Biol.* **8**, 367–406.

STEWARD, F. C. and PRESTON, C. (1941). The effect of salt concentration upon the metabolism of potato disks and the contrasted effect of potassium and calcium salts which have a common ion. *Pl. Physiol.* **16**, 85–116.

STEWARD, F. C. and STREET, H. E. (1947). The nitrogenous constituents of plants. *A. Rev. Biochem.* **16**, 471–502.

STIEHLER, R. D. and FLEXNER, L. B. (1938). A mechanism of secretion in the chorioid plexus: the conversion of oxidation–reduction energy into work. *J. biol. Chem.* **126**, 603–17.

SUTCLIFFE, J. F. (1960). New evidence for a relationship between ion absorption and protein turnover in plant cells. *Nature, Lond.* **188**, 294–7.

SUTCLIFFE, J. F. (1962). *Mineral Salt Absorption in Plants*. Oxford, London New York, Paris: Pergamon Press.

TAGER, J. M., VELDSEMA-CURRIE, R. D. and SLATER, E. C. (1966). Chemiosmotic theory of oxidative phosphorylation. *Nature, Lond.* **212**, 376–9.

TORII, K. and LATIES, G. G. (1966). Dual mechanisms of ion uptake in relation to vacuolation in corn roots. *Pl. Physiol.* **41**, 863–70.

Ussing, H. H. (1949). Transport of ions across cellular membranes. *Physiol. Rev.* **29**, 127–55.

Ussing, H. H. (1957). General principles and theories of membrane transport. In *Metabolic Aspects of Transport Across Cell Membranes*, pp. 39–56. Madison: University of Wisconsin Press.

Van Lookeren Campagne (1957). The action spectrum of the influence of light on chloride absorption in *Vallisneria* leaves. *Proc. K. ned. Akad. Wet.* **60**, 70–6.

Vasington, F. D. and Murphy, J. V. (1961). Active binding of calcium by mitochondria. *Fedn. Proc.* **20**, 146.

Vasington, F. D. and Murphy, J. V. (1962). Ca^{++} uptake by rat kidney mitochondria and its dependence on respiration and phosphorylation. *J. biol. Chem.* **237**, 2670–7.

Walker, N. A. (1957). Ion permeability and the plasmalemma of the plant cell. *Nature, Lond.* **180**, 94.

Whittam, R. (1962). The asymmetrical stimulation of a membrane adenosine triphosphatase in relation to active cation transport. *Biochem. J.* **84**, 110–18.

Whittam, R. and Ager, M. E. (1965). The connexion between active cation transport and metabolism in erythrocytes. *Biochem. J.* **97**, 214–27.

Williams, R. J. P. (1961). Possible functions of chains of catalysts. *J. Theoret. Biol.* **1**, 1–17.

Williams, R. J. P. (1962). Possible functions of chains of catalysts II. *J. Theoret. Biol.* **3**, 209–29.

INDEX

95